Selected Titles in This Series

(*Continued in the back of this publication*)

Single Orbit Dynamics

Conference Board of the Mathematical Sciences

CBMS

Regional Conference Series in Mathematics

Number 95

Single Orbit Dynamics

Benjamin Weiss

Published for the
Conference Board of the Mathematical Sciences
by the
American Mathematical Society
Providence, Rhode Island
with support from the
National Science Foundation

CBMS Conference on Probabilistic Aspects of Single Orbit Dynamics
held at California State University, Bakersfield
June 1995

Research partially supported by the
National Science Foundation

1991 *Mathematics Subject Classification*. Primary 22D40, 28Dxx, 60Gxx.

Library of Congress Cataloging-in-Publication Data
Weiss, Benjamin, 1941–
    Single orbit dynamics / Benjamin Weiss.
        p. cm. — (Regional conference series in mathematics, ISSN 0160-7642 ; no. 95)
    Includes bibliographical references.
    ISBN 0-8218-0414-6 (alk. paper)
    1. Ergodic theory—Congresses.   2. Orbit method—Congresses.   3. Stochastic processes—
Congresses.   I. National Science Foundation (U.S.)   II. Title.   III. Series.
QA1.R33   no. 95
[QA611.5]
510 s—dc21

[515′.42]
                                                                                           99-050183

Contents

Preface

These notes represent a mildly expanded version of a series of ten lectures that I gave at a CBMS conference organized by Kamel Haddad at California State University, Bakersfield, CA in June, 1995. Due to external circumstances their publication has been delayed for a few years but I hope that they still give a timely presentation of a novel point of view in dynamical systems. I have not made any effort to update the exposition, but I would like to point out some recent developments that are relevant to the reader who wishes to pursue matters further.

First of all I would like to recommend the new book by Paul Shields, *The ergodic theory of discrete sample paths*, Grad. Studies in Math. v. 13 (AMS), 1996. In this book there is a very careful treatment of some central issues in ergodic theory and information from a point of view that is close to that expounded in these lectures. There is a particularly good treatment there of entropy related matters and of various characterizations of Bernoulli processes.

Following up on an idea proposed by M. Gromov, Elon Lindenstrauss and I have developed a new invariant in topological dynamics which refines the classical notion of topological entropy. This invariant, called the mean topological dimension, vanishes for all systems with finite topological entropy but distinguishes between various systems with infinite topological entropy. There is a single orbit interpretation of this invariant which should shed some new light on spaces of meromorphic functions, solutions of dynamical systems with infinitely many degrees of freedom etc. The basic theory is set out in a joint paper *Mean Topological Dimension*, (to appear in the Israel J. of Math).

The style of these notes is that of a lecture. When proofs are given they are meant to be complete, but not every i is dotted nor every t crossed. Most of the material appears elsewhere and I have given references at the end of each chapter to guide the reader who wants to pursue matters in more detail. Chapter 4 contains results that haven't appeared before in print. They originate in discussions that I had with Don Ornstein fifteen years ago when we were traveling on a weekly basis from Stanford to MSRI. The main result in chapter 5 is due to Y. Katznelson, I thank him for his permission to include it in these lectures. Hillel Furstenberg's influence on these lectures began with a course that I took with him thirty five years ago at Princeton during which he gave the first exposition of his ideas on disjointness which he then called absolutely independent. It has continued ever since and culminated in a careful reading that he, and Eli Glasner, gave of these notes. Naturally the responsibility for all remaining errors is mine alone.

Finally I would like to express my thanks to Kamel Haddad who organized the wonderful conference that made these notes possible. Last, but not least my thanks to Stanford University and Willene Perez, who typed there the first draft of these notes, and to Shani Ben David, the Hebrew University of Jerusalem, who is responsible for the final version.

Benjamin Weiss, July 1999

Jerusalem (TVBBA), Israel

CHAPTER 1

What is Single Orbit Dynamics

The study of dynamical systems has its origins in the classical mechanics of
Newton and his successors. The modern theory has flowered in several different
directions that are best described by abstracting certain features of the classical
systems. The ergodic theorem of J. von Neumann pointed the way to the functional analytic - **operator ergodic theory** while the seminal ergodic theorem of
G.D. Birkhoff founded what we now call **measure ergodic theory** which can be
viewed as the study of transformations of measure spaces. Keeping only the topological structure led to the **topological dynamics** of Gottschalk and Hedlund and
focusing on the analytic or smooth structures gave rise to **complex dynamics**
and **smooth dynamics**. Each of the above has expanded to a broad and fairly
independent discipline with its own problems and methodology. Less well known
specialties of a similar nature are **generic dynamics** where first category, in the
sense of Baire, replaces the μ-null sets as the negligible sets, and **measurable or
Borel dynamics** where no sets are neglected a priori and the only structure that
comes into play is the Borel structure.

Single orbit dynamics is not another kind of dynamics of the type described
above. It is rather an attempt to focus attention on a rather large body of work
concerned with dynamical study of single orbits as opposed to the global study of
a system as a whole. One can see the origins of this point of view in the works of
R. von Mises, H. Bohr and N. Wiener none of whom actually thought about what
they were doing in the terms that I just used. Nonetheless, I will begin by telling
something about the relevant work of each of these mathematicians before I try to
give some general formulation of single orbit dynamics.

At the beginning of the century probability theory was not yet established as a bona fide member of the community of mathematics. Its status was closer to that of a physical theory where the phenomena that were being described were games of chance and other random events and processes. Richard von Mises attempted to base a mathematical theory of probability on the primitive notion of a **collective**. This was supposed to capture the idea of a random sequence of outcomes (H-heads or T-tails) of a simple experiment such as tossing a coin. His axiomatic description of such a sequence of simple coin tossing was basically:

    I. The asymptotic frequency of occurrences of H in the collective equals $1/2$.

    II. Property I persists for any subsequence of outcomes derived from the collective by place selection rule.

In order to avoid misrepresenting his ideas here is a direct quote from an English version of his book Probability, Statistics and Truth posthumously published by his widow Hilda Geiringer:

*A collective appropriate for the application of the theory of probability must fulfill two conditions. First, the relative frequencies of the attributes must possess limiting values. Second, these limiting values must remain the same in all partial sequences which may be selected from the original one in an arbitrary way. Of course, only such partial sequences can be taken into consideration as can be extended indefinitely, in the same way as the original sequence itself. Examples of this kind are, for instance, the partial sequences formed by all odd members of the original sequence, or by all members for which the place number in the sequence is the square of an integer, or a prime number, or a number selected according to some other rule, whatever it may be. The only essential condition is that the question whether or not a certain member of the original sequence belongs to the selected partial sequence should be settled **independently of the result** of the corresponding observation, i.e., before anything is known about this result. We shall call a selection of this kind a **place selection**. The limiting values of the relative frequencies in a collective must be independent of all possible place selections. By place selection we mean the selection of a partial sequence in such a way that* *we decide whether an element should or should not be included without making use of the attribute of the element, i.e., the result of our game of chance.*

While this may appear strange to generations of mathematicians who have been raised on Kolmogorov's treatment of probability theory as a special kind of measure theory it represented an earlier, systematic mathematical foundation of probability. What is important for us is his insistence that one can calculate probabilities from a **single** sequence of outcomes.

During the 1930's the approach of von Mises fell into disfavor, after the criticisms of Tornier, Ville, Frechet and others who pointed out the difficulties in establishing the existence of collectives. The criticisms hinge, of course, on what interpretation one places on the phrase "admissible place selections". We shall discuss these matters in more detail in a later lecture after we will have some ergodic theoretic tools at our disposal.

Next we turn to H. Bohr's theory of almost periodic functions. Motivated by his attempts to prove the famous Riemann hypotheses on the zeros of the zeta function he was led to consider bounded functions on the reals $f(t)$ that generalized the usual periodic functions. Here is his definition:

DEFINITION 1.1. A complex valued bounded function on $\mathbb{R}$, $f(t)$ is **almost periodic** if for any $\varepsilon > 0$, the numbers $p \in \mathbb{Z}$ that satisfy

$$(1) \qquad\qquad \sup_{t \aleph} |f(t) - f(t+p)| \leq \epsilon$$

are **syndetic**, i.e. the gaps between successive p's are bounded.

Any p which satisfies (1) is called an almost period. Bohr showed that just like periodic functions can be expanded in a Fourier series so too the almost periodic functions have an expansion as a generalized Fourier series. More precisely we have:

THEOREM 1.2. *If f is almost periodic then there is a countable set $\Lambda \subset \mathbb{R}$ such that for all $\epsilon > 0$ there is an exponential polynomial*

$$Q(t) = \sum_{j=1}^{J} c_i e^{i\lambda_j t}, \qquad \{\lambda_1, \dots, \lambda_J\} \subset \Lambda$$

that satisfies

$$\sup_t |Q(t) - f(t)| \leq \epsilon.$$

This class of functions and many of its generalizations became an active field of investigation and turned out to have many applications. We mention one later result due to Favard:

THEOREM 1.3. *If $f(t)$ is a bounded function such that*

$$f(t+1) - f(t) = g(t)$$

is almost periodic then $f(t)$ is also an almost periodic function.

On the face of it there seems to be no connection between this set of ideas and dynamical systems. Nonetheless, S. Bochner pioneered an approach to the theory of almost periodic functions which fits perfectly into the framework of topological dynamics and in particular to what we mean by single orbit dynamics. This will be explained in detail in the next lecture.

Our third historical example is the generalized harmonic analysis of N. Wiener. Motivated by problems in the newly emerging field of electrical engineering and communication theory he was led to define the **spectrum** of a function $\xi(n)$ as the set of complex numbers σ of modulus one such that

$$\lim_{N\to\infty} \frac{1}{2N+1} \sum_{-N}^{N} \bar{\sigma}^n \xi(n) \neq 0.$$

Naturally for sequences to have a spectrum it is necessary that this limit exist for all σ with $|\sigma| = 1$. As he later showed, stationary stochastic processes, provide a rich source for such sequences. We shall discuss these matters in the third lecture, for now it suffices for us to see here another example of some kind of nontrivial analysis of a single sequence being the center of a series of investigations.

All three of these examples, the collectives of von Mises, the almost periodic functions of Bohr and the time series of Wiener illustrate how individual functions are important objects of investigations. As we will see, all three of these examples can be very profitably studied using certain concepts and techniques drawn from dynamical systems. This is one aspect of single orbit dynamics. Here are some more recent examples of this type that center around the general theme of patterns in large sets. The most famous of this class of results is the theorem of van der

Waerden that in any finite partition of the integers at least one of the sets contains arithmetic progressions of arbitrary length.

About twenty years ago, H. Furstenberg and his collaborators showed how ideas from dynamical systems could be used to obtain old and new results of this type. The basic paradigm involved interpreting the object under investigation as a single orbit of an entire dynamical system. This study of patterns in large sets has developed into a broad discipline and has injected powerful new methods into basic combinatorial questions. I shall describe in a little more detail, now, a more recent example of this which we shall discuss in detail in Chapter 5.

For any set D of positive integers we define a graph $G(D)$ with vertex set $\mathbb{Z}$ and edges consisting of all pairs (i, j) such that $|i - j|$ belongs to D. This $G(D)$ is a translation invariant graph on the integers. A very natural question is: when does $G(D)$ have a finite chromatic number? This question can be translated into a question of topological dynamics and then some standard methods there lead, for example, to the result that whenever D is **lacunary** $G(D)$ has a finite chromatic number. On the other hand, for any growth rate that is less than lacunary there are sequences D with that growth rate for which the chromatic number of $G(D)$ is infinite.

In all of the above we are dealing with the **existence** of patterns. It turns out that the **rate** of recurrence of patterns can also be examined using dynamical tools. If $x = x_1 x_2 x_3 \ldots$ is an infinite sequence on a finite alphabet A, we can define

$$R_n(x) = \min\{r > n : x_{r+i} = x_i \qquad 1 \le i \le n\}$$

if there is such an r, and set $R_n(x) = \infty$ otherwise. For regular sequences x that are generated by stationary stochastic processes it turns out that $R_n(x)$ has a well defined growth rate that is intimately related to Shannon's entropy. This connection makes it possible to give sense to the information content of individual sequences.

Up to now the emphasis has been on that aspect of single orbit dynamics in which the main object of interest is a single orbit and its properties and the global dynamics is a tool. The reverse situation, in which the main object of interest is the global system and the individual orbit is but a tool is also not uncommon. We call this the principle of **one for all**. In topological dynamics this means describing a global system as the orbit closure of a single orbit. This is a powerful tool in the construction of a variety of examples. A classical example is the Morse minimal sequence defined as follow: our alphabet will be $\{0, 1\}$, and we agree that $\bar{0} = 1$, $\bar{1} = 0$. If B is a block consisting of k-symbols then $\bar{B}$ is the k-block obtained by replacing each $b \in B$ by $\bar{b}$. Define inductively:

$$B_0 = 0,$$

$$\vdots$$

$$B_{n+1} = B_n \bar{B}_n$$

$$\vdots$$

then $B_n \to x = x_0 x_1 x_2 \ldots$ which is called the Morse minimal sequence. It begins as follows

$$x = 0110100110010110 \ldots$$

and its orbit closure in $\{0, 1\}^{\mathbb{N}}$ under the shift map provides a very basic example in topological dynamics.

In ergodic theory orbit closures alone are not sufficient and one adds the notion of a **generic point** which allows one to define probabilities in terms of a single orbit. Statistical investigations in which one tries to learn the global features of a system from an individual sample sequence also are covered by this principle of **one for all** in the probabilistic framework. We will study several examples of this in greater detail below. These include questions like the classical ones:

- How do the successive observations $\xi_1 \xi_2 \xi_3 \ldots$ of a stationary stochastic process enable one to give a better and better description of the process (as n tends to infinity)?

- How well can one predict the next output ξ_1 as one learns more and more about the past $\xi_0 \xi_{-1} \xi_{-2} \ldots$?

At this point I would like to give a concrete example of this single orbit philosophy in action. A very classical result in probability theory is the recurrence of the one dimensional simple random walk. Let $\{x_n\}_1^{\infty}$ be independent random variables with distribution

$$P(x_n = +1) = P(x_n = -1) = 1/2 \qquad \text{for all } n,$$

and define

$$S_n = x_1 + x_2 + \cdots + x_n.$$

The recurrence of this simple random walk is the statement that with probability one S_n will take the value 0. It is known that the same holds true if the increments x_n come from any stationary stochastic process with zero mean. Not so well known is the following generalization.

THEOREM 1.4. *Let $\{x_n\}$ be a $\mathbb{Z}$-valued stationary stochastic process such that for all $\epsilon > 0$*

$$P\left\{\left|\frac{1}{n}\sum_1^n x_i\right| > \epsilon\right\} \to 0 \qquad \text{as } n \to \infty.$$

Then with probability one $S_n = 0$.

Even for independent random variables this is a stronger result than the usual one since the hypothesis may be satisfied even when $E\{|x_i|\}$ is infinite.

Denote by Ω the underlying probability space of the random variables x_i, and fix a small $\epsilon > 0$. Let

$$E_n = \left\{\omega : \left|\frac{1}{n}\sum_1^n x_i(\omega)\right| \le \epsilon^2\right\}$$

and let N_0 be sufficiently large so that for all n larger than N_0, $P(E_n) > 1 - \frac{\epsilon^2}{2}$, and then let N be much larger than N_0 $\left(\text{say } N > \left(\frac{2}{\epsilon^2}N_0\right)\right)$. Consider now

$$f(\omega) = \frac{1}{N}\sum_1^N 1_{E_n}(\omega).$$

We have arranged matters so that the integral of f is greater than $1 - \epsilon^2$, and since $0 \le f \le 1$ it follows that on most of the space $f(\omega) \ge 1 - \epsilon$.

We focus now on a single ω for which $f(\omega) \ge 1 - \epsilon$ and consider the graph of

$$S_n(\omega) = \sum_1^n x_i(\omega)$$

thought of as a function from $\{1, 2, \ldots, N\}$ to $\mathbb{Z}$. For most of the n's we have that $|S_n(\omega)| \le \epsilon N$, and thus only a small fraction of the indices n_0 can have the property that the value $S_{n_0}(\omega)$ is not repeated for some $n_1 > n_0$. Thus if we view the progress of the random walk along this typical ω, it turns out that for most of our visits $S_{n_0}(\omega)$ that site is visited again at a later time $n_1 \leq N$.

Up to this point we haven't used the stationarity of the increment process. Now this stationarity will be used to translate the recurrence at later times to recurrence to the origin - 0 - our position when the walk begins. We again use an averaging argument. For a set with probability at least $1 - \epsilon$ we know that $1 - 2\epsilon$ of the indices $n_0 \leq N$ have the property that there is an $n_0 < n_1 \leq N$ for which

$$S_{n_1}(\omega) = S_{n_0}(\omega).$$

It follows that for some **fixed** n_0, there is a set of ω of probability at least $1 - 3\epsilon$ so that for each of these ω's there is an $n_1 > n_0$ with $S_{n_1}(\omega) - S_{n_0}(\omega) = 0$. What this means is that if we would define a random walk with increments $x_{n_0+1}, x_{n_0+2}, \ldots$ then this one would return to the origin with probability at least $1 - 3\epsilon$. But by stationarity this is the same as the probability that the original random walk returns to the origin. Since ϵ was arbitrary this completes the proof of the theorem. $\square$

The proof of the theorem takes place by analyzing closely the behavior of a single orbit $S_n(\omega)$, with "Fubini's theorem", or "averaging in different ways" being used to translate the global hypothesis, the weak law of large numbers for the x_n's, into information concerning a single orbit. This type of interplay between the individual orbits and the global properties will be a recurring theme in the rest of these lectures.

The final point that I would like to make in this introductory lecture is the distinction between almost everywhere results and single orbit theorems. I will illustrate this with two examples - one drawn from von Mises collectives and the other from Wiener's generalized harmonic analysis.

DEFINITION 1.5. A number $x \in (0,1)$ is normal in the base b if in the b-ary expansion of x

$$x = \frac{\xi_1}{b} + \frac{\xi_1}{b^2} + \cdots + \frac{\xi_i}{b^i} + \cdots \xi_i \in \{0, 1, \ldots, b-1\}$$

any block $B \in \{0, 1, \ldots, b-1\}^K$ occurs with an asymptotic frequency equal to b^{-K}, $(K = 1, 2, \ldots)$.

More formally the condition means that for all such B;

$$\lim_{n\to\infty} \frac{1}{n} |\{i \le n : \xi_i \xi_{i+1} \cdots \xi_{i+k-1} = B\}| = b^{-K}.$$

At the beginning of the century, E. Borel proved that the set of non-normal numbers has Lebesgue measure zero. It is an easy consequence of his proof that for any sequence $n_1 < n_2 < \ldots$ a.e. x has the property that $\xi_{n_1} \xi_{n_2} \xi_{n_3} \ldots$ is also a normal number. These are examples of a.e. theorems. The following theorem characterizes those positive density sequences $\{n_j\}$ that have the property that $\xi_{n_1} \xi_{n_2} \ldots$ is normal whenever x is normal. This is a single orbit theorem since it asserts something about individual orbits and conclusions can be drawn about specific points. For the formulation of the theorem a preliminary definition is necessary.

Let us denote by Ω the space $\{0,1\}^{\mathbb{N}}$ with the product topology and let $T : \Omega \to \Omega$ denote the shift on Ω, so that $(T\omega)(n) = \omega(n+1)$. Probability measures on Ω that are T-invariant define $\{0,1\}$-valued stationary stochastic processes (the coordinates $x_n(\omega) = \omega(n)$) and vice versa. Such a process is called **deterministic** if x_0 is measurable with respect to the σ-field generated by the $\{x_i : i \ge 1\}$. The topology on the space of T-invariant measures that we consider is the w^*-topology which is such that $\mu_n \to \mu$ if for all finite cylinder sets C $\mu_n(C)$ converges to $\mu(C)$.

DEFINITION 1.6. A sequence $u \in \{0,1\}^{\mathbb{N}}$ is said to be **completely deterministic** if any invariant probability measures that is in the closure of the set of measures:

$$\nu_N = \frac{1}{N} \sum_{n=1}^{N} \delta_{T_u^n}$$

defines a deterministic process.

THEOREM 1.7 (T. Kamae - B. Weiss). *A positive density sequence* $\{n_1 < n_2 < \ldots\}$ *has the property that for every normal number*

$$x = .\xi_1 \xi_2 \xi_3 \ldots$$

also $.\xi_{n_1} \xi_{n_2} \xi_{n_3} \ldots$ *is normal if and only if the indicator function of the sequence is completely deterministic.*

The indicator function of the sequence $\{n_j\}$ is the element $u \in \{0,1\}^{\mathbb{N}}$ such that $u(n) = 1$ if and only if n is one of the n_j's. Simple examples of completely deterministic sequences are $[j\alpha]$ for any real $\alpha \geq 1$. Note that there are uncountably many distinct completely deterministic sequences.

A generalization of the notion of a normal number is the following. If T is a continuous mapping of a compact metric space X to itself and μ is a T-invariant probability measure on X then $x_0 \in X$ is said to be generic for the system (X, T, μ) if for all continuous functions f on X we have

$$\lim_{n \to \infty} \frac{1}{n} \sum_{i=1}^{n} f(T^i x_0) = \int_X f(x) d\mu(x).$$

For the coordinate shift on Ω above, being a normal number coincides with being a generic point for the product measure

$$\prod_0^{\infty} \left(\frac{1}{2}, \frac{1}{2} \right) \quad \text{on} \quad \Omega$$

with respective to T, the coordinate shift.

Here is a result of Wiener-Winter that applies to **all** generic points of a system.

THEOREM 1.8. *If x_0 is generic for (X, T, μ) and the only eigenfunctions that T has as an operator on $L^2(X, \mu)$ are the constants then for all complex λ of modulus one $\lambda \neq 1$ and all continuous functions*

$$\lim_{N \to \infty} \frac{1}{N} \sum_1^{N} \lambda^j f(T^j x_0) = 0.$$

This is the single orbit result from the generalized harmonic analysis of N. Wiener. It provides a rich class of examples of bounded sequences for which a spectral analysis is possible. One way to use such a result is to observe that the failure of the limit above to vanish for some λ means that the system has some point spectrum.

To sum up briefly, we have discussed several kinds of results involving single orbits — ranging from the analysis of individual sequences through the construction of entire systems from single orbits up to conclusions that are valid for all generic points of certain systems. These results all point to the centrality of the behavior of individual orbits and to highlight this focus we call this body of work single orbit dynamics. In the following lectures we will illustrate all of these themes by concrete examples.

References

1. G.D. Birkhoff, *Dynamical systems*, AMS Colloq. 9, Providence, R.I., 1927.
2. H. Bohr, *Almost periodic functions*, Chelsea, New York, 1951.
3. J. Favard, *Sur les equations differentielles lineaires à coefficients presque periodiques*, Acta Math. **51**(1928), 31–81.
4. H. Furstenberg, *Stationary processes and prediction theory*, Annals of Math Studies 44, Princeton, N.J., 1960.
5. H. Furstenberg, *Recurrence in Ergodic Theory and Combinatorial Number Theory*, Princeton, N.J., 1981.
6. W.H. Gottschalk and G.A. Hedlund, *Topological Dynamics*, AMS Colloq. 36, Providence, R.I., 1955.
7. T. Kamae, *Subsequences of normal sequences*, Israel J. of Math. **16**(1973), 121–149.
8. Richard von Mises, *Probability Statistics and Truth*, revised English edition by Hilda Geiringer, London, NY, 1957.
9. B. Weiss, Normal Sequences as Collectives, in *Proceeding of Symposium on Topological Dynamics and Ergodic Theory*, Univ. of Kentucky, 1971.
10. B. Weiss and T. Kamae, *Normal Numbers and selection rules*, Israel J. of Math. **21**(1975), 101–110; Israel J. of Math. **21**(1975), 159–166.
11. B. Weiss, *Measurable Dynamics*, in Conference in Modern Analysis and Probability, Cont. Math. **25**(1984), 395–421.
12. B. Weiss, D. Sullivan and J. Wright, *Generic dynamics and monotone complete C^*-algebras*, TAMS, **295**(1986), 795–809.
13. B. Weiss and D. Ornstein, *How sampling reveals a process*, Annals of Prob. **18**(1990), 905–930.
14. B. Weiss, H. Furstenberg and Y. Katznelson, *Ergodic theory and configurations in sets of positive density*, in Mathematics of Ramsey Theory, ed. by J. Nesetril and V. Rodl, Springer, 1990, pp. 184–199.
15. N. Wiener, *Generalized harmonic analysis*, Acta Math. **55**(1930), 117–258.

CHAPTER 2

Topological Dynamics

Briefly put, topological dynamics is the study of the iterates of a continuous mapping of a topological space. We will confine our attention to compact metric spaces (X, ρ) and will often, but not always, require that the continuous mapping $T : X \rightarrow X$ be a homeomorphism. The simplest infinite example is obtained abstractly by compactifying $\mathbb{Z}$ with the addition of one point $\{\infty\}$ and defining T by:

$$T(n) = n + 1, \qquad T(\infty) = \infty.$$

This basic example occurs whenever one has a fixed point for T in X, say $Tx_\infty = x_\infty$ and an infinite orbit which is attracted to x_∞ both in positive time and negative time – say an $x_0 \in X$ such that $T^n x_0 \rightarrow x_\infty$ as $|n| \rightarrow \infty$, but $T^n x_0 \neq x_0$ for all $n \neq 0$. There is a simple general construction which when applied to this basic example will give us one of our fundamental, nontrivial, examples.

The basic construction can be thought of as a generalization of finite Cartesian products – when these are viewed as the action of T on finite ordered subsets of X. Let us denote by 2^X the **closed subsets** of X. We put the **Hausdorff metric** on 2^X defined as follows: Denote by A_ϵ the set

$$A_\epsilon = \{y \in X : \text{ for some } a \in A, \ \rho(a, y) < \epsilon\}$$

and set

$$\rho(A, B) = \inf\{\epsilon : A \subset B_\epsilon \text{ and } B \subset A_\epsilon\}.$$

With this metric, 2^X becomes a compact space, and if T is a homeomorphism of X, T induces a homeomorphism of 2^X via the natural action $TA = \{Ta : a \in A\}$.

If (X_0, T_0) denotes our previous example, $X_0 = \mathbb{Z} \cup \{\infty\}$ with $T_0(n) = n + 1$, then observe that 2^{X_0} consists of two parts – the finite subsets of $\mathbb{Z}$, and arbitrary subsets of $\mathbb{Z}$ together with $\{\infty\}$. This latter part, which forms the bulk of $2^{\aleph_0}$, can be identified with $\{0,\ 1\}^{\mathbb{Z}}$ with the product topology and the shift transformation S defined by $(Sx)(n) = x(n+1)$, for $x \in \{0,\ 1\}^{\mathbb{Z}}$. This is sometimes called the 2-shift and will serve us as one of our basic examples of the general theory. This transformation, or its noninvertible factor, the shift on $\{0,\ 1\}^{\mathbb{N}}$, occurs in many concrete examples and is the home of the simplest **symbolic dynamics**. Consider, for example, the iterations of the mapping $z \to z^2$ of the complex plane to itself. This transformation has two attracting fixed points – zero and infinity and nontrivial dynamics takes place only on the unit circle $e^{2\pi it}$. Here the transformation takes the form $t \to 2t(\mathrm{mod}\ 1)$. This is easily visualized as its graph consists of two straight lines of slope 2. Consider now the dyadic representation of number $t \in [0,\ 1]$:

$$t = \sum_1^{\infty} d_j(t)/2^j, \qquad d_j \in \{0,\ 1\}.$$

In this representation the mapping $t \to 2t$ becomes the shift on the sequence of digits

$$(d_1,\ d_2,\ d_3,\dots\ ) \longrightarrow (d_2,\ d_3,\dots\ \ \ ).$$

The advantages of this representation are that the entire orbit is visualized at a glance – one can see, together with the point, all of its images under successive iterations of the mapping. Topologically, the space of the shift is completely disconnected – so that we do not have the circle in a precise way. What we do have is a continuous mapping π,

$$\pi : \{0,\ 1\}^{\mathbb{N}} \to [0,\ 1]$$

given by $\pi(x) = \sum_9^{\infty} x(n)/2^n$ which is one to one at all but a countable number of points (the dyadic rationals) where π is two to one. This mapping π takes the shift to multiplication by two.

In general, if we have two systems $(X_1,\ T_1)$, $(X_2,\ T_2)$ we say that $(X_2,\ T_2)$ is a **factor** of $(X_1,\ T_1)$ or that $(X_1,\ T_1)$ is an **extension** of $(X_2,\ T_2)$ if there is a continuous surjective mapping $\pi : X_1 \to X_2$ such that for all $x_1 \in X_1$

$$\pi(T_1(x_1)) = T_2(\pi(x_1)).$$

Looking inside a system (X, T), if $E \subset X$ is closed and T invariant then we say that (E, T) is a **subsystem** of (X, T). Here is another basic example, the Hilbert cube $\Omega = [0, 1]^{\mathbb{Z}}$ with T equal to the coordinate shift. Now it is easy to manufacture factors of an arbitrary system (X, T) in the following way. Start with any continuous function $f : X \to [0, 1]$ and define

$$F : X \to \Omega$$

by $F(x) = (\dots f(T^{-1}x),\ f(x),\ f(T_x),\dots f(T^n x),\dots)$ and look at $F(X)$. This is closed and shift invariant and is therefore a subsystem of (Ω, T) and it is clearly a factor of (X, T). With these basic categorical-type operations we can formulate a question:

QUESTION 2.1. *Is every metric system (X, ρ, T) a subsystem of $([0, 1]^{\mathbb{Z}}, \text{shift})$?*

The meaning, of course, is can f be chosen so that the corresponding F is one to one? Now there is an obvious obstruction which comes about from the looking at the fixed points and periodic points of (Ω, T). Namely the fixed points of (Ω, T) are homeomorphic to the unit interval, so it is clear that the identity transformation on a space not imbeddable in $[0, 1]$ cannot be imbedded in (Ω, T). To avoid such trivialities let us assume that (X, T) has no periodic points at all. Then in full generality the question is open. A. Jaworski proved some years ago that the answer is positive if the topological dimension of the space X is finite.

The most fundamental dynamical systems are the **minimal** ones. These are systems (X, T) which have no nontrivial subsystems. Since any orbit closure, $\overline{\{T^n x\}_{-\infty}^{\infty}}$ is closed, T-invariant and nonempty, an equivalent condition is that the orbit of every point is dense. The basic fact discovered by G. D. Birkhoff is that any system (X, T) has minimal subsystems (E, T). The sets E such that (E, T) is minimal are often called the minimal sets of (X, T). Usually one proves this by appealing to Zorn's lemma to give the existence of minimal elements in the collection of closed T-invariant subsets of X. However, one can give a more constructive demonstration as follows: List a basis for the topology of X as a sequence of open sets $U_1, U_2, \dots$. Consider $\bigcup\limits_{-\infty}^{\infty} T^i U_1$; if it is **not** all of X then set

$$X_1 = X \backslash \bigcup_{-\infty}^{\infty} T^i \, U_1,$$

if it is all of X set $X_1 = X$. In either case X_1 is T-invariant. To the extent that $U_1 \cap X_1$ is nonempty we have that the orbit of every point in X_1 intersects U_1. We continue with $\tilde{U}_2 = U_2 \cap X_1$, namely set $X_2 = X_1$ if $\bigcup_{-\infty}^{\infty} T^i \, U_2 \supset X_1$ and otherwise set $X_2 = X_1 \backslash \bigcup_{-\infty}^{\infty} T^i \, U_2$. Continuing in this fashion we set finally $E = \bigcap_{n=1}^{\infty} X_n$. Since X is compact, E is nonempty, and for each U_k that intersects E at all we have $\bigcup_{-\infty}^{\infty} T^i \, U_k \supset E$, and since $U_k \cap E$ is a basis for the topology of E we see that every point in E has a dense orbit. One can view this procedure as an attempt to find a point with a dense orbit. Any point in a minimal set E will be called a **minimal point**. Thus $x_0 \in X$ is a minimal point if any y in the orbit closure of x_0 has x_0 in its own orbit closure, $\overline{\{T^n y\}}$. As a byproduct we get that any compact system has **recurrent** points, that is points x_0 that satisfy

$$\liminf_{n \to \infty} \rho(T^n x_0, \, x_0) = 0.$$

These are more general than minimal points. It is important to note that the fact that all orbits are dense comes together with a uniformity, since for compact spaces if $\bigcup_{-\infty}^{\infty} T^i U = X$ then already for some finite N, $\bigcup_{-N}^{N} T^i U = X$ or $\bigcup_{0}^{2N} T^i U = X$. Another way to say this is to say that the visit times of any point $x \in X$ to an open set U form a **syndetic** sequence in $\mathbb{Z}$, that is, a sequence with bounded gaps.

Classical examples of minimal systems are

(i) finite periodic orbits,

(ii) irrational rotations of the circle $|z| = 1$, i.e. the mapping taking $e^{i\theta}$ to $e^{i(\theta + \alpha)}$ with $\alpha/2\pi$ irrational, and

(iii) the horocycle flow on a compact surface with strictly negative curvature. For a self-contained definition of this flow one can say the following: Let Γ be a co-compact lattice in $S\ell(2, \mathbb{R})$, and let $X = S\ell(2, \mathbb{R})/\Gamma$, then the horocycle flows are the flows defined by the unipotent groups $\left\{ \begin{pmatrix} 1 & 0 \\ t & 1 \end{pmatrix} \right\}$, or $\left\{ \begin{pmatrix} 1 & t \\ 0 & 1 \end{pmatrix} \right\}$ acting on X.

Let me reassure the reader for whom this is not familiar that I will not be saying anything later on about this example. I merely wanted to point out that there are classically-defined systems that are minimal. The minimality of (i) is obvious, that of (ii) is a theorem of Kronecker that is easily proved using the fact that any infinite subset of $\{|z| = 1\}$ must have some cluster point. The minimality of the horocycle flow in (iii) is not so easily proved, it was done some sixty years ago by G. Hedlund.

Let us return now to the almost-periodic functions of Bohr. Recall that a real valued function f on $\mathbb{Z}$ is **almost periodic** if for every $\epsilon > 0$, the set of ϵ-almost periods is syndetic. Explicitly, for every $\epsilon > 0$, there is some M such that any interval $(a, a + M)$ contains an L such that $|f(n + L) - f(n)| < \epsilon$ for all n. Denote by B the set of bounded functions with the distance

$$\|f - g\| = \sup_n |f(n) - g(n)|, \qquad f, g \in B.$$

Define $T : B \to B$ by $(Tb)(n) = b(n + 1)$, $n \in \mathbb{Z}$ and observe that T is an **isometry**.

It is well known that B is a complete metric space in the metric defined above. Let X denote the closure of the orbit of f, an almost periodic function, under T. The defining property of almost periodicity is exactly equivalent to the assertion that X is compact! Indeed the orbit $\{T^j f\}_{j \in \mathbb{Z}}$ is totally bounded, since given ϵ, if M is as in the definition then for any translate $T^a f$, look at $[a - M, a]$. This contains an ϵ-translation number, say ℓ, and if $u = a - \ell$ we have:

$$\|f - T^\ell f\| < \epsilon$$

whence $\|T^u f - T^u(T^\ell f)\| < \epsilon$ since T is an isometry and so $\|T^u f - T^a f\| < \epsilon$. Thus $\{f, Tf, \ldots T^M f\}$ is an ϵ-spanning subset of the orbit.

Starting with an almost-periodic function we have constructed a compact dynamical system (X, ρ, T), where T is an isometry for some metric ρ on X, and there is $x_0 \in X$ with dense orbit. We proceed to analyze such systems – which are in a sense the simplest type. Equip the set of isometries of X, $I(X)$ with a metric ρ defined by

$$\rho(T_1, T_2) = \sup_x \rho(T_1 x, T_2 x)$$

and note that in the definition one can calculate the supremum over any dense set, and in particular over the orbit $\{T^n x_0\}_{n \in \mathbb{Z}}$. This easily implies that if $T^{n_i} x_0$

converges to a point y_0 then $\{T^{n_i}\}$ converges in $I(X)$ to an isometry S, satisfying

$$Sx_0 = y_0$$

and indeed if G denotes the closure of $\{T^n\}$ in $I(X)$ then G is a compact group which is identified with X by mapping y_0 to S.

Coming back to our almost-periodic function f, let G denote the compact group that we have constructed, so that f, the base point x_0 in the abstract construction, is identified with e, the identity of G. Evaluation at 0, defines a **continuous function** $\tilde{f}$ on G, and f is recovered as $f(n) = \tilde{f}(T^n)$. The fundamental theorems of Bohr's theory can be obtained now from the elementary theory of characters defined on compact commutative groups. This approach was introduced by S. Bochner and represented a great conceptual simplification over the original "bare hands" proofs given by H. Bohr.

As a further illustration of this approach let us see how H. Furstenberg proved a classical theorem of Favard with these dynamical ideas.

THEOREM 2.2. *If $g(n)$ is a bounded complex valued function on $\mathbb{Z}$ such that $g(n+1) - g(n)$ is Bohr almost periodic then so is g.*

The proof of this "single-function" theorem is based on a dynamical theorem of W. Gottschalk and G. Hedlund:

THEOREM 2.3. *If $(X,\ T)$ is minimal and $f\ :\ X \to \mathbb{C}$ is continuous and satisfies*

$$\left| \sum_{i=0}^{M} f(T^i x) \right| \le b$$

for some b and all $x \in X$, $M \in \mathbb{N}$, then there is a continuous function $h: X \to \mathbb{C}$ such that

$$f(x) = h(Tx) - h(x).$$

PROOF. Consider $\widehat{T} :\ X \times \mathbb{C} \to X \times \mathbb{C}$ defined by

$$\widehat{T}(x,\ z) = (Tx,\ z + f(x)).$$

Fix $x_0 \in X$ and let Y be the orbit closure of $(x_0,\ 0)$ under $\widehat{T}$. This set is bounded, by hypothesis, and hence is compact. It has therefore a minimal subset, say E. Since $(X,\ T)$ was minimal the projection of E onto the first coordinate is all of

X. For any $\omega_0 \in \mathbb{C}$, the mapping $S_{\omega_0}(x, z) = (x, z + \omega_0)$ commutes with $\widehat{T}$ and therefore takes a compact minimal set for $\widehat{T}$ into a compact minimal set. If there would be two points in E over a single $x \in X$, say

$$(x, z_1), (x, z_2) \in E$$

set $\omega_0 = z_2 - z_1$. Then $S_{\omega_0} E \cap E \neq \emptyset$, and since E is minimal that would mean that $S_{\omega_0} E = E$. But clearly the orbit under S_{ω_0} is not bounded and this is a contradiction. We conclude that E has only a single point over each $x \in X$, and thus is defined by a function $h : X \to \mathbb{C}$, $E = \{(x, h(x)) : x \in X\}$. Since E is closed h is continuous and finally the $\widehat{T}$ invariance of E implies

$$h(Tx) = h(x) + f(x)$$

which gives the theorem.   □

To prove Favard's theorem we begin now with the almost-periodic function $f(n) = g(n + 1) - g(n)$. Using Bochner's construction that we described above, we see that f is the restriction of a continuous function on a compact group K to a copy of $\mathbb{Z}$ defined by some element $\gamma \in K$, so that $\mathbb{Z}\gamma$ is dense in K, and the system (K, γ) is minimal!

Now the representation of f as $g(n + 1) - g(n)$, gives us that f satisfies the hypothesis of the G-H theorem and thus we get a continuous function h on K such that

$$f(k) = h(k + \gamma) - h(k)$$

and thus when restricting to $\mathbb{Z}\gamma$ we see that

$$g(n + 1) - g(n) = h(n + 1) - h(n).$$

This gives that $g - h$ is constant on $\mathbb{Z}\gamma$, and thus it follows that g and h differ by a constant. Since h is continuous on K, its restriction to $\mathbb{Z}\gamma$ is almost periodic and therefore so is $g(n)$ which yields theorem 2.

Next to the isometries the simplest systems are the **distal** ones. The most natural way to describe them relies on the following generalization of minimality:

DEFINITION 2.4. A system is said to be **semisimple** or **pointwise minimal** if every $x \in X$ belongs to a minimal set.

It is not so easy to say something about general semisimple systems since they include all minimal systems, for example. However, it turns out that there is an alternative characterization of those systems (X, T) for which the Cartesian product $(X \times X, T \otimes T)$ is semisimple and this is **distality**. A metric system (X, ρ, T) is said to be **distal** if for all $x \neq y$ one has

$$\inf_n \rho(T^n x, T^n y) > 0.$$

For nonmetric systems there is an analogous definition in terms of the uniform structure on X. The basic proposition is:

PROPOSITION 2.5 (R. Ellis). *A system is distal if and only if $(X \times X, T \otimes T)$ is semisimple.*

For the proof of this proposition we need to introduce one more notion. We say that x and y are **proximal** if

$$\inf_n \rho(T^n x, T^n y) = 0.$$

Even if x does not belong to a minimal set there is always some minimal set in the ω-limit set of x, namely those points that are proper cluster points of $\{T^n x\}_{n \geq 0}$. More is true, namely we have the fundamental:

LEMMA. *Any point x in a compact system (X, T) is proximal to some minimal point y.*

Thus not only are there minimal points in the ω-limit set of x, but there are minimal points y in the ω-limit set that are tracked along some $n_i \to \infty$ by x in the sense that

$$\lim_{n_i \to \infty} \rho(T^{n_i} x, T^{n_i} y) = 0.$$

For the proof of the lemma some more machinery is needed and we shall simply accept it – referring for the proof to [Fu-1981] for example. From the lemma we get a corollary which is stronger than one of the implications in the proposition.

COROLLARY 2.6. *Any distal system is semisimple.*

PROOF. For any x, we know that x is proximal to some minimal point y. Since $(X,\ T)$ is distal there are no nontrivial proximal pairs so we conclude that $y = x$, i.e. that x itself is a minimal point which is the assertion of the corollary.   □

PROOF OF THE PROPOSITION 2.5. If $(X,\ T)$ is distal then clearly so is $X \times X$ distal, and therefore it is semisimple by the corollary. Conversely, suppose $X \times X$ is semisimple and that $(x,\ y)$ is a proximal pair. Then $(x,\ y)$ belongs to a minimal set $E \subset X \times X$, under $T \times T$, but as a proximal pair, the orbit closure meets the diagonal which is surely invariant under $T \times T$ and thus E must coincide with the diagonal and $x = y$ and the nontrivial proximal pair does not exist.   □

Here is a family of examples of distal transformations. Start with any distal transformation $(X,\ T)$ and let $\phi :\ X \to \mathbb{R}$ be continuous. Define

$$T_\phi :\ X \times \mathbb{T}^1 \longrightarrow X \times \mathbb{T}^1$$

by the formula

$$T_\phi(x,\ \eta) = (Tx,\ \eta + \phi(x)) \text{ where the addition is modulo 1.}$$

In particular, if T is an irrational rotation of $\mathbb{T}^1$ you get the map sending

$$(\zeta, \eta) \to (\zeta + \alpha,\ \eta + \phi(\zeta)).$$

For a special case consider the map

$$(\zeta, \eta) \longrightarrow (\zeta + 2\alpha,\ \eta + \zeta + \alpha),$$

the nth iterate has the form $(\zeta + n\alpha,\ \eta + n\zeta + n^2\alpha)$, and so information about sequences like $n^2\alpha$ can be extracted from knowledge of the system as a whole.

In conclusion I would like to discuss briefly a rather recent generalization of distality. Let us say that a $(X,\ \rho,\ T)$ is **mean distal** if for any $x \neq y$

$$\overline{\lim_{N \to \infty}} \frac{1}{2N+1} \sum_{-N}^{N} \rho(T^n x,\ T^n y) > 0.$$

The idea here is that instead of requiring that for all n, the distances between $T^n x$ and $T^n y$ should remain greater than some fixed constant, we are prepared to ignore what happens along a subsequence of density zero. To formulate the main results that have been gotten so far (in collaboration with D. Ornstein and J. King) about this class I need the notion of **topological entropy**. This will be defined and studied in detail later on (Chapter VII). For the moment it will suffice to say that the topological entropy $h_{\text{top}}(X, T)$ is a nonnegative number, including $+\infty$, associated to a system which is invariant under topological conjugacy. Topological conjugacy between systems (X_i, T_i) is a homeomorphism θ between X_1 and X_2 such that $T_2\theta = \theta T_1$. The entropy measures in a sense how much randomness there is in a system. For the k-shift, which is the full shift on k-symbols, $\{1, 2, \ldots k\}^{\mathbb{Z}}$, the topological entropy is $\log k$, while for any distal system the topological entropy is zero. The following theorem was obtained together with D. Ornstein:

THEOREM 2.7. *If (X, T) is mean distal and has finite topological entropy then its topological entropy must be zero.*

Surprisingly enough the following question is open:

QUESTION 2.8. *Do infinite entropy mean distal transformations exist?*

Turning to structural issues, it is easy to see that a factor of a distal system is distal. This follows from the fact that a factor of a minimal system is minimal which implies that a factor of a semisimple system is semisimple. For mean distal systems we have no such nice characterization and indeed the situation is much more complicated. First of all one can see that not all zero-entropy systems are mean distal. While the classical minimal zero entropy nondistal system – the horocycle flow – is mean distal, one can fairly easily construct many other examples which are not, for example the system defined by the orbit closure of $\text{sign}(\cos n\alpha)$ in $\{-1, +1\}^{\mathbb{Z}}$, for irrational $\alpha/2\pi$ is not mean distal. There one can find distinct points that are even doubly asymptotic. On the other hand we have established, with J. King, the following:

THEOREM 2.9. *Any zero-entropy subsystem of $\{0, 1\}^{\mathbb{Z}}$ has a mean distal extension.*

The proof of this result is a good illustration of the single-orbit method since the construction is carried out by writing down a second symbolic sequence directly above the orbits of the original system. It would take us too far afield to go into any more details at this point.

References

1. G.D. Birkhoff, *Dynamical Systems*, AMS - revised edition, 1966.
2. H. Furstenberg, *Stationary Processes and Prediction Theorem*, Princeton Univ. Press, 1960.
3. H. Furstenberg, *Recurrence in Ergodic Theorem and Combinatorial Number Theory*, Princeton Univ. Press, 1981.

CHAPTER 3

Invariant Measures, Ergodicity and Unique Ergodicity

Probability theory will enter the picture in the guise of probability measures on X that are invariant under the action of T. In general, there need exist no such measure, e.g. $\mathbb{Z}$ with $T(n) = n + 1$, and much effort has gone into the study of what kind of conditions will guarantee the existence of invariant measures. The simplest condition that ensures this is compactness and throughout this lecture X will be a compact space. To see that invariant measures always exist in this case we set $\mathcal{P}(X)$ equal to the set of probability measures on X and provide $\mathcal{P}$ with a topology so that it becomes a **compact** space. This is an abstract formulation of an old theorem due to Helly and Bray. For those versed in functional analysis it suffices to say that we view $\mathcal{P}(X)$ as a subset of $C(X)^*$, where $C(X)$ is the space of continuous functions on X with the uniform norm, and then use the weak * topology on $C(X)^*$. It is easily verified that $\mathcal{P}(X)$ is a closed subset of the unit ball in $C(X)^*$ and thus Alaoglu's theorem gives the required compactness.

More explicitly, for compact metric spaces, we can proceed as follows: Let $\{f_j\}_{j=1}^{\infty}$ denote a countable family of positive continuous functions $\{f_j\}_1^{\infty}$ of norm one ($\sup_x |f_j(x)| = 1$, $j = 1, 2, \ldots$) whose linear span is dense in $C(X)$. Define a metric on $\mathcal{P}(X)$ by setting

$$d(\mu, \nu) = \sum_1^{\infty} 2^{-j} \left| \int_X f_j(x) \, d\mu(x) - \int_X f_j(x) \, d\nu(x) \right|.$$

Convergence of measures μ_n in this metric is the pointwise convergence of the numerical sequences $\{\int f_j d\mu_n\}$ for each j. The Riesz theorem identifies probability measures on X with positive linear functionals on $C(X)$ that give the value 1 to the constant function **one** on X. We can clearly assume $f_1 \equiv 1$, and now the compactness of $\mathcal{P}(x)$ follows from the usual compactness of bounded sequences in the topology of pointwise convergence, i.e. the compactness of the Hilbert cube $[0,1]^{\mathbb{N}}$ in the product topology. We shall usually suppress the σ-algebra on which the measures are defined but understand always that it is the σ-algebra generated by the topology on X, namely the Borel sets $\mathcal{B}$. Notice finally that, in the topology we have defined, a sequence μ_n converges to μ if and only if for all continuous functions f we have

$$\lim_{n \to \infty} \int_X f(x) \, d\mu_n(x) = \int_X f(x) \, d\mu(x).$$

Now starting with any measure μ, we can form the sequence $\mu_n = \frac{1}{n} \sum_{j=0}^{n-1} T^j \mu$ where $T^j \mu$ is defined by

$$\int_X f(x) \, d(T^j \mu)\,(x) = \int_X f(T^j x) \, d\mu(x)$$

and then any cluster point of the sequence $\{\mu_n\}$ is an invariant measure. This latter fact follows at once from the observation that for all bounded f:

$$\left| \int_X f(x) \, d\mu_n(x) - \int_X f(Tx) \, d\mu_n(x) \right| \le \frac{2}{n} \, \|f\|.$$

In particular, we can start with a Dirac measure at the point x_0 and then we get some subsequence $n_i \to \infty$ such that for all $f \in C(X)$

$$(*) \qquad \lim_{i \to \infty} \frac{1}{n_i} \sum_{k=0}^{n_i-1} f(T^k x_0) = \int_X f(x) \, d\mu(x)$$

with μ being an invariant measure under T. If $(*)$ holds for $n_i = i$ we say that the point x_0 is **generic** for the measure μ. There are systems, such as the full shift on $A^{\mathbb{Z}}$, where **all** invariant measures can be obtained in this way. In any event, when x_0 is generic for an invariant measure μ we can think of the single orbit $\{T^n x_0\}$ as representing the probability space $(X, \mathcal{B}, \mu)$. Putting it another way, the frequencies defined by the observations $f(T^n x_0)$ yield the average value of f with respect to μ.

This is the natural place to define **ergodic measures**. In general, for a measure space $(Y, \mathcal{C}, \nu)$ and a measure preserving transformation $T : Y \to Y$, i.e. one for which $T^{-1}C \in \mathcal{C}$ for all $C \in \mathcal{C}$ and $\nu(T^{-1}C) = \nu(C)$ or equivalently:

$$\int_Y f(y)\, d\nu(y) = \int_Y f(Ty)\, d\nu(y) \qquad \text{all } f \in L^1(Y,\, \mathcal{C}, \nu)$$

we say that $(Y,\, \mathcal{C},\, \nu,\, T)$ is **ergodic** if $\nu(T^{-1}C \triangle C) = 0$ implies that either $\nu(C) = 0$ or $\nu(Y \backslash C) = 0$. If the transformation is fixed we will say that ν is an **ergodic measure**. In the situation above, it is straightforward to verify that T-invariant measures form a **closed convex** subset of $\mathcal{P}(X)$, which we denote by $\mathcal{I}(X)$, and that the ergodic measures are precisely the extreme points of this set. Indeed, if $\mu \in \mathcal{I}(X,\, T)$ and μ is not ergodic, and $B \in \mathcal{B}$ is an invariant Borel set such that $0 < \mu(B) < 1$, then $\frac{1}{\mu(B)} \cdot \mu|_B$ also defines an element of $\mathcal{I}(X,\, T)$ and

$$\mu = \mu(B) \left(\frac{1}{\mu(B)}\, \mu|_B \right) + \mu(X \backslash B) \left(\frac{1}{\mu(X \backslash B)}\, \mu|_{(X \backslash B)} \right)$$

shows that μ is **not** an extreme point of $\mathcal{I}(X,\, T)$. Conversely, if μ is not an extreme point, and

$$\mu = \frac{1}{2}\, \lambda_1 + \frac{1}{2}\, \lambda_2 \text{ with } \lambda_1,\, \lambda_2 \in \mathcal{I}(X,\, T)$$

then λ_1 is absolutely continuous with respect to μ and the Radon-Nikodym derivative $\frac{d\lambda_1}{d\mu}$ is a nonconstant **invariant** function. For some level u, the set $\{x \in X : \frac{d\lambda_1}{d\mu}(x) < u\}$ will give a nontrivial invariant set.

The well-known Krein-Milman theorem gives now the existence of ergodic measures and the general theory of compact convex sets in a locally convex metrizable space gives that any invariant measure can be represented as an integral of ergodic ones. Even without the general theory if $\mathcal{I}(X)$ consists of a single point then the above discussion shows that this measure must be ergodic. In this case we call the system $(X,\, T,\, \mu)$ **uniquely ergodic**. If, in addition, $\text{supp}(\mu) = X$ we will say that the system is **strictly ergodic**. Clearly $\text{supp}(\mu)$, the closed support of X, is a closed invariant subset of X and thus a uniquely ergodic system is strictly ergodic when restricted to $\text{supp}(\mu)$. We connect this discussion with our previous discussion of minimality in:

PROPOSITION 3.1. *A strictly ergodic system $(X,\, T,\, \mu)$ is minimal.*

PROOF. If $E \subset X$ would be a nontrivial closed invariant subset then by the discussion above the system $(E,\, T)$ has an invariant measure ν with support contained in E. This ν is not μ since μ has global support and thus we have contradicted the strict ergodicity. □

This is the place where we should discuss generic points in general, and for this we need the Birkhoff ergodic theorem. Since proofs of this abound we will merely formulate the classical theorems of von-Neumann and Birkhoff for later reference.

MEAN ERGODIC THEOREM. *If* $(X, \mathcal{B}, \mu, T)$ *is a measure preserving system and* $I_0 \subset L_2(X, \mathcal{B}, \mu)$ *is the closed subspace of* T*-invariant functions with* π_0 *the projection onto it, then for all* $f \in L_2$

$$\lim_{N \to \infty} \left\| \frac{1}{N} \sum_0^{N-1} f(T^n x) - (\pi_0 f)(x) \right\|_{L_2} = 0.$$

Ergodicity is then equivalent to the fact that I_0 consists of the constants.

INDIVIDUAL ERGODIC THEOREM. *If* $(X, \mathcal{B}, \mu, T)$ *is a measure-preserving system and* π_0 *represents the projection of* $L_1(X, \mathcal{B}, \mu)$ *onto the* T*-invariant integrable functions then for all* $f \in L_1$ *and a.e.* x

$$\lim_{N \to \infty} \frac{1}{N} \sum_0^{N-1} f(T^n x) = (\pi_0 f)(x).$$

Taking a countable dense set of functions in $C(X)$, for an **ergodic** measure where $\pi_0 f$ is simply the integral, one sees that generic points always exist – in fact almost every point is generic.

As a concrete example, take the group K of complex numbers of modulus one, and $\rho = e^{2\pi i \alpha}$ with α irrational, and define R_ρ to be multiplication by ρ. More generally let G be a metric compact monothetic group such that $\{\rho^n : n \in \mathbb{Z}\}$ is dense in G we can look at the system (G, R_ρ) where $R_\rho(g) = \rho g$. According to our discussion above there is a probability measure μ on G invariant under R_ρ. Since for any $\beta \in G$, there is a sequence ρ^{n_i} that converges to β we have for any continuous function f

$$\int_G f(\beta g)\, d\mu(g) = \int_G \lim f(\rho^{n_i} g)\, d\mu(G)$$

$$= \lim \int_G f(\rho^{n_i} g)\, d\mu(G)$$

$$= \int_G f(g)\, d\mu(G)$$

where the interchange between limits and integral is justified by the uniform convergence of $f(\rho^{n_i} g)$ to $f(\beta g)$. Notice that G must be commutative and then if ν is any other such invariant measure we could evaluate

$$\int_G \int_G f(g_1\, g_2)\, d\mu(g_1)\, d\nu(g_2)$$

in two different ways by Fubini's theorem and get that $\mu = \nu$. (This is the usual proof of the uniqueness of Haar measure for compact abelian groups.) We have just shown that **minimal group rotation systems are always strictly ergodic.** These are also called **Kronecker systems** because Kronecker proved a higher dimensional analogue of the fact that for irrational α, $n\alpha$ is dense modulo 1.

Let us return now to our generic points. Recall that in a compact metric space if a sequence has a unique cluster point then it converges to that point. Applying this observation to the construction of invariant measures above proves the following:

PROPOSITION 3.2. *If $(X,\, T,\, \mu)$ is strictly ergodic then every point $x_0 \in X$ is generic for μ.*

This establishes a famous theorem of H. Weyl on the equidistribution in $[0,\, 1]$ of the fractional parts of $n\alpha$, when α is irrational. Indeed, as we have seen for irrational α, the sequence of fractional parts of $n\alpha$ is dense in $[0,\, 1]$ and thus Lebesgue measure is the unique measure invariant under addition by α (mod 1). Now the fact that 0 is a generic point for this measure under this transformation means that for f a continuous function with period 1

$$(3.1) \qquad \lim_{N \to \infty} \frac{1}{N} \sum_{1}^{N} f(n\alpha) = \int_0^1 f(t)dt$$

which easily gives Weyl's theorem on equidistribution. That is usually formulated just like (3.1) but using indicator functions of intervals. However, it is easy to see that if (3.1) holds for all continuous functions then it also holds for all Riemann integrable functions such as the indicator functions of intervals.

Yet another way of characterizing unique ergodicity is an immediate consequence of the Hahn-Banach theorem. A measure μ on X is T-invariant if and only if it vanishes on all functions of the form $T \circ f - f$. It follows, therefore, that this is unique if and only if the **uniform closure** of $\{T \circ f - f : f \in C(X)\}$ is of codimension one in $C(X)$. Now observe that for functions g which can be written as $T \circ f - f$ we have

$$\frac{1}{N} \sum_{0}^{N-1} g(T^n x) = \frac{1}{N}(f(T^N x) - f(x))$$

and thus these averages converge uniformly to zero. Clearly this extends to the uniform closure and since it is trivially true for the constants we have established:

PROPOSITION 3.3. *The system (X, T, μ) is uniquely ergodic if and only if for every continuous function f*

$$\lim_{N \to \infty} \left\| \frac{1}{N} \sum_{0}^{N-1} f(T^n x) - \int_X f(x) \, d\mu(x) \right\|_\infty = 0.$$

One can give a relative version of these ideas as follows: Suppose that (X, T) is a factor of (Y, S) with a continuous factor mapping $\pi : Y \to X$. Let μ be an invariant measure for T such that there is a **unique** measure λ with $\pi \circ \lambda = \mu$. That there is always **some** such invariant measure can be seen as follows: The mapping π imbeds $C(X)$ as a subspace in $C(Y)$. Thus by the Hahn-Banach theorem we can extend μ to some linear functional on $C(Y)$. The fact that the norm is 1 and that the value of the functional on 1 is 1 guarantees the positivity and thus μ extends to a probability measure on Y. The set of measures on Y that project onto μ is seen to be a closed T-invariant (since $T\mu = \mu$) subset of $\mathcal{P}(Y)$ and by the same argument that showed that there were always T-invariant measures we see that this set contains an S-invariant measure. Repeating the extreme point analysis we see that there are also ergodic measures over μ in case μ was ergodic. We call the above situation, when the extension is unique, a **uniquely ergodic extension**. Repeating the proof of the proposition (3.2) in this context gives:

PROPOSITION 3.4. *If* (Y, S, λ) *is a uniquely ergodic extension of* (X, T, μ) *then for any* $x_0 \in X$ *that is generic for* μ, *every* $y_0 \in \pi^{-1}(x_0)$ *is generic for* λ.

Here is an application of these ideas. Consider a compact system (X, T) with an ergodic invariant measure μ. Let λ be a measure on

$$Z = X \times \mathbb{T}^1$$

invariant under S, defined by

$$S(x, \zeta) = (Tx, \ e^{2\pi i \alpha} \cdot \zeta)$$

ergodic and projecting onto μ. For each β, let R_β denote the mapping defined by

$$R_\beta(x, \zeta) = (x, \ e^{2\pi i \beta}\zeta).$$

The β's such that $R_\beta \lambda = \lambda$ form a closed subgroup H_0. Here are the possibilities for this subgroup:

A) H_0 is all of $\mathbb{T}^1$. Then $\lambda = \mu \times$Lebesgue measure will be the **unique** measure over μ, i.e. this extension will be uniquely ergodic. To see this suppose that ν is some other S-invariant measure over μ. Then we have in any event $R_\beta \nu$ is also S-invariant for any β and

$$\lambda = \int\limits_0^1 R_\beta \circ \nu d\beta$$

follows from Fubini's theorem. Since λ is ergodic the integral representation must be trivial, i.e. $R_\beta \nu$ must equal λ for almost every β which easily gives that $\nu = \lambda$.

Now for **every** generic point x_0, we get that all points above it, in particular 1, are generic for λ and applying this to the function $e^{2\pi i t} f(x)$, for $f \in C(X)$ we get for $\eta = e^{2\pi i \alpha}$

$$\lim_{N \to \infty} \frac{1}{N} \sum_0^{N-1} \eta^n \ f(T^n x_0) = 0.$$

B) For no nonzero β is it the case that $R_\beta \lambda = \lambda$. Now looking at generic points for λ we see that if y_0 is generic for λ then $R_\beta y_0$ is generic for $R_\beta \lambda$. It follows that a single fiber over a $x_0 \in X$ can contain at most one point that is generic for λ. Since the set of generic points for λ has full measure (here we use the individual ergodic theorem) we see that the generic points define $\mu - a.e.$ a function ψ from X to $\mathbb{T}^1$. The invariance of the set of generic points under S gives now that the function ψ satisfies the equation

$$\psi(Tx) = e^{2\pi i \alpha} \psi(x) \qquad \mu - a.e.$$

and thus $e^{2\pi i \alpha}$ is in the point spectrum of T.

C) H_0 is finite, say the kth roots of unity. In this case if we map $\mathbb{T}^1$ to itself by sending ζ to ζ^k we can return to case B), and our conclusion is that for some k, $e^{2\pi i k \alpha}$ is an eigenvalue of T.

If we collect this together we have established a version of the Wiener-Wintner theorem which we formulate as follows:

THEOREM 3.5. *If no power of $e^{2\pi i \alpha}$ is an eigenvalue of the ergodic transformation (X, T, μ) then the unique invariant measure on $(X \times \mathbb{T}^1, T \times R_\alpha)$ is $\mu \times$ Lebesgue measure, and in that case for every generic point x_0 and every continuous function f we have*

$$\lim_{N \to \infty} \frac{1}{N} \sum_0^{N-1} e^{2\pi i n \alpha} f(T^n x_0) = 0.$$

Since at most a countable number of α's can be eigenvalues for an ergodic system (X, β, μ, T) the same arguments applied to any L^1-function f give the existence of a set of full μ-measure of x_0's for which the limit in theorem 3.5 exists for **all** α. This is the usual statement of the Wiener-Wintner theorem. However, 3.5 contains more precise information in that it says that the limit holds for any generic point.

It came as quite a surprise when Jewett showed that any weakly mixing transformation can be modeled by a uniquely ergodic system. Until then the uniquely ergodic systems were still viewed as rather special and the examples that were known had a limited scope. Shortly afterwards, W. Krieger showed that the restriction on weak mixing could be removed and he established:

THEOREM 3.6. *Any ergodic* $(Y, \mathcal{C}, \nu, S)$ *has a uniquely ergodic system* (X, T, μ) *that is isomorphic to it.*

By isomorphism we mean, of course, a measurable invertible mapping $\theta : Y \to X$ that takes ν to μ and intertwines S with T, i.e. $T\theta = \theta S$. Several proofs have been given of this theorem and I would like to sketch now an approach which will serve as a pattern for our work in the next chapter. A key notion is that of a **uniform** partition whose importance in this context was emphasized by G. Hansel and J.P. Raoult.

DEFINITION 3.7. A set $B \in \mathcal{B}$ is uniform if

$$\lim_{N \to \infty} \ \text{ess} \sup_{x} \ \left| \frac{1}{N} \sum_{0}^{N-1} 1_B(T^i x) - \mu(B) \right| = 0.$$

A partition $\mathcal{P}$ is uniform if, for all N, every set in $\bigvee_{-N}^{N} T^{-i}\mathcal{P}$ is uniform.

The connection between uniform sets, partitions and unique ergodicity lies in proposition 3.3. It follows easily from that proposition that if $\mathcal{P}$ is a uniform partition, say into the sets $\{P_1, P_2, \ldots P_a\}$, and we denote by $\mathcal{P}$ also the mapping that assigns to $x \in X$, the index $1 \le i \le a$ such that $x \in P_i$, then we can map X to $\{1, 2, \ldots a\}^{\mathbb{Z}} = A^{\mathbb{Z}}$ by:

$$\pi(x) = (\ldots \mathcal{P}(T^{-1}x), \ \mathcal{P}(x), \ \mathcal{P}(Tx), \ldots \mathcal{P}(T^n x), \ldots).$$

Pushing forward the measure μ by π, gives us a measure $\pi \circ \mu$ on $A^{\mathbb{Z}}$, and the closed support of this measure will be a closed shift invariant subset, say $E \subset A^{\mathbb{Z}}$. Now the indicator functions of finite cylinder sets span the continuous functions on E, and the fact that $\mathcal{P}$ is a uniform partition and proposition 3.3 combine to establish that (E, shift) is uniquely ergodic. This will not be a model for $(X, \mathcal{B}, \mu, T)$ unless $\bigvee_{-\alpha}^{\infty} T^{-i}\mathcal{P} = \mathcal{B}$ modulo null sets, but in any case this does give a model for a nontrivial factor of X.

Our strategy for proving theorem 3.6 is to begin by constructing a single nontrivial uniform partition. Then this partition will be refined more and more via uniform partitions until we generate the entire σ-algebra $\mathcal{B}$. Along the way we will be showing how one can prove a relative version of the basic Jewett–Krieger theorem. Our main tool is the use of Rohlin towers. These are sets $B \in \mathcal{B}$ such that for some N, B, TB, ... $T^{N-1}B$ are disjoint while $\bigcup_{0}^{N-1} T^i B$ fill up most of the space. Actually we need Kakutani–Rohlin towers, which are like Rohlin towers but fill up the whole space. If the transformation does not have rational numbers in its point spectrum this is not possible with a single height, but two heights that are relatively prime, like N and $N+1$ are certainly possible. Here is one way of doing this. The ergodicity of $(X, \mathcal{B}, \mu, T)$ with μ non atomic easily yields, for any n, the existence of a positive measure set B, such that

$$T^i \, B \cap B = \emptyset, \qquad i = 1, \, 2, \ldots n.$$

With N given, choose $n \geq 10 \cdot N^2$ and find B that satisfies the above. It follows that the return time

$$r_B(x) = \inf\{i > 0 : T^i x \in B\}$$

is greater than $10 \cdot N^2$ on B. Let

$$B_\ell = \{x \in B : r_B(x) = \ell\}.$$

Since ℓ is large (if B_ℓ is nonempty) one can write ℓ as a positive combination of N and $N+1$, say

$$\ell = N u_\ell + (N+1) v_\ell.$$

Now divide the column of sets $\{T^i B_\ell : 0 \leq i < \ell\}$ into u_ℓ-blocks of size N and v_ℓ-blocks of size $N+1$ and mark the first layer of each of these blocks as belonging to C. Taking the union of these marked levels ($T^i B_\ell$ for suitably chosen i) over the various columns gives us a set C such that r_C takes only two values – either N or $N+1$ as required.

It will be important for us to have at our disposal K-R towers like this such that the columns of say the second K-R tower are composed of entire subcolumns of the earlier one. More precisely we want the base C_2 to be a subset of C_1 – the base of the first tower. Although we are not sure that this can be done with just two column heights we can guarantee a bound on the number of columns that depends only on the maximum height of the first tower. Let us define formally:

DEFINITION 3.8. A set C will be called the base of a bounded K-R tower if for some N, $\bigcup_{0}^{N-1} T^i C = X$ up to a μ-null set. The least N that satisfies this will be called the height of C, and partitioning C into sets of constancy of r_C and viewing the whole space X as a tower over C will be called the K-R tower with columns the sets $\{T^i C_\ell : \ 0 \leq i < \ell\}$ for $C_\ell = \{x \in C : r_C(x) = \ell\}$.

Our basic lemma for nesting these K-R towers is:

LEMMA 3.9. *Given a bounded K-R tower with base C and height N, for any n sufficiently large there is a bounded K-R tower with base D contained in C whose column heights are all at least n and at most $n + 4N$.*

PROOF. We take an auxiliary set B such that $T^i \, B \cap B = \emptyset$ for all $0 < i < 10(n + 2N)^2$ and look at the unbounded (in general) K-R tower over B. Using the ergodicity it is easy to arrange that $B \subset C$. Now let us look at a single column over B_m, with $m \geq 10 \, (n + 2N)^2$. We try to put down blocks of size $n + 2N$ and $n + 2N + 1$, to fill up the tower. This can certainly be done but we want our levels to belong to C. We can refine the column over B_m into a finite number of columns so that each level is either entirely within C or in $X \backslash C$. This is done by partitioning the base C according to the finite partition:

$$\bigcap_{i=0}^{m-1} T^{-i}\{C, \ X \backslash C\}.$$

Then we move the edge of each block to the nearest level that belongs to C. The fact that the height of C is N means that we do not have to move any level more than $N - 1$ steps, and so at most we lose $2N - 2$ or gain that much. Thus our blocks, with bases now all in C, have a size in the interval $[n, \ n + 4N]$ as required, and we can define D to be the union of the bases. $\qquad\square$

It is clear that this procedure can be iterated to give an infinite sequence of nested K-R towers with a good control on the variation in the heights of the columns. These can be used to construct uniform partitions in a pretty straightforward way, but we need one more lemma which strengthens slightly the ergodic theorem. We will want to know that when we look at a bounded K-R tower with base C and with minimum column height sufficiently large that for most of the fibers of the towers (that is for $x \in C$, $\{T^i x : \ 0 \leq i < r_C(x)\}$) the ergodic averages of some finite set of functions are close to the integrals of the functions. It would seem that there is a problem because the base of the tower is a set of very small measure (less than 1/min column height) and it may be that the ergodic theorem is not valid there. However, a simple averaging argument using an intermediate size gets around this problem. Here is the result which we formulate for simplicity for a single function f:

LEMMA 3.10. *Let f be a bounded function and $(X, \mathcal{B}, \mu, T)$ ergodic. Given $\epsilon > 0$, there is an n_0, such that if a bounded K-R tower with base C has minimum column height at least n_0, then those fibers over $x \in C$: $\{T^i x : 0 \le i < r_C(x)\}$ that satisfy*

$$\left| \frac{1}{r_C(x)} \sum_{i=0}^{r_C(x)-1} f(T^i x) - \int_X f d\mu \right| < \epsilon$$

fill up at least $1 - \epsilon$ of the space.

PROOF. Assume without loss of generality that $|f| < 1$. For a δ to be specified later find an N such that the set of $y \in X$ which satisfy

$$(*) \qquad \left| \frac{1}{N} \sum_0^{N-1} f(T^i y) - \int f d\mu \right| < \delta$$

has measure at least $1 - \delta$. Let us denote the set of y that satisfy $(*)$ by E. Suppose now that n_0 is large enough so that N/n_0 is negligible – say at most δ. Consider a bounded K-R tower with base C and with minimum column height greater than n_0. For each fiber of this tower, let us ask what is the fraction of its points that lie in E. Those fibers with at least a $\sqrt{\delta}$ fraction of its points not in E cannot fill up more than a $\sqrt{\delta}$ fraction of the space, because $\mu(E) > 1 - \delta$.

Those fibers with more than a $(1 - \sqrt{\delta})$ fraction of its points lying in E we can divide now into disjoint blocks of size N that cover all points that lie in E by moving up the fiber, marking the first point in E, skipping N steps and continuing until we exhaust the height of the fiber. If we choose $\sqrt{\delta}$ to be less than $\epsilon/10$, then we can now get that on most of the fiber the averages over those N-blocks is close to $\int f d\mu$ and since $|f| \le 1$ its values off of these good N-blocks cannot spoil this result so that indeed we have established the lemma. $\square$

We are now prepared to construct uniform partitions. Start with some fixed nontrivial partition $\mathcal{P}_0$. By lemma 3.10, for any tall enough bounded K-R tower at least $9/10$ of the columns will have the 1-block distribution of each $\mathcal{P}_0$-name within $\frac{1}{10}$ of the actual distribution. We build a bounded K-R tower with base $C_1(1)$ and heights N_1, $N_1 + 1$ with N_1 large enough for this to be valid. It is clear that we can modify $\mathcal{P}_0$ to $\mathcal{P}_1$ on the bad fibers so that now all fibers have a distribution of 1-blocks within $\frac{1}{10}$ of a fixed distribution. We call this new partition $\mathcal{P}_1$. Our further changes in $\mathcal{P}_1$ will not change the N_1, $N_1 + 1$ blocks that we see on fibers of a tower over our ultimate C_1. Therefore, we will get a uniformity (up to $\frac{1}{10}$) on all blocks of size $100N_1$. The 100 is to get rid of the edge effects since we only know the distribution across fibers over points in $C_1(1)$.

Next we apply lemma 3.10 to the 2-blocks in $\mathcal{P}_1$ with $1/100$. We choose N_2 so large that N_1/N_2 is negligible and so that any bounded K-R tower with height at least N_2 has for at least $99/100$ of its fibers a distribution of 2-blocks within $1/100$ of the global $\mathcal{P}_1$ distribution. Find now a bounded K-R tower with base $C_2(2) \subset C_1(1)$ such that its column heights are between N_2 and $N_2 + 4N_1$. For the fibers with good $\mathcal{P}_1$ distribution we make no change. For the others, we copy on most of the fiber (except for the top $10 \cdot N_1^2$ levels) the corresponding $\mathcal{P}_1$-name from one of the good columns. In this copying we also copy the $C_1(1)$-name so that we preserve the blocks. The final $10 \cdot N_1^2$ spaces are filled in with N_1, $N_1 + 1$ blocks. This gives us a new base for the first tower that we call $C_1(2)$, and a new partition $\mathcal{P}_2$. The features of $\mathcal{P}_2$ are that all its fibers over $C_1(2)$ have good (up to $1/10$) 1-block distribution, and all its fibers over $C_2(2)$ have good (up to $1/100$) 2-block distributions. These will not change in the subsequent steps of the construction.

Note too that the change in $C_1(1)$, to $C_1(2)$, could have been made arbitrarily small by choosing N_2 sufficiently tall. With this done it is clear how one continues to build a sequence of partitions $\mathcal{P}_n$ that converge to $\mathcal{P}$ and $C_i(k) \to C_i$ such that the $\mathcal{P}$-names of all fibers over points in C_i have a good (up to $1/10^i$) distribution of i-blocks. This clearly gives the uniformity of the partition $\mathcal{P}$ as required and establishes

PROPOSITION 3.11. *Given any $\mathcal{P}_0$ and any $\epsilon > 0$ there is a uniform partition $\mathcal{P}$ such that $d(\mathcal{P}_0, \mathcal{P}) < \epsilon$ in the ℓ_1-metric on partitions.*

REMARK 3.12. We skipped over one point in the above proof and that is the filling in of the top relatively small portion of the bad fibers after copying over most of a good fiber. We cannot copy an exact good fiber because it is conceivable that no fiber with the precise height of the bad fiber is good. The filling in is possible if the column heights of the previous level are relatively prime. This is automatically the case if there is no rational spectrum. If there are only a finite number of rational points in the spectrum then we could have made our original columns with heights LN_1, $L(N_1 + 1)$ with L being the highest power so that T^L is not ergodic and then worked with multiples of L all the time. If the rational spectrum is infinite then we get an infinite group rotation factor and this gives us the required uniform partition without any further work.

As we have already remarked the uniform partition that we have constructed gives us a uniquely ergodic model for the factor system generated by this partition. We need now a relativized version of the construction we have just carried out. We formulate this as follows:

PROPOSITION 3.13. *Given a uniform partition $\mathcal{P}$ and an arbitrary partition $\mathcal{Q}_0$ that refines $\mathcal{P}$, for any $\epsilon > 0$ there is a uniform partition $\mathcal{Q}$ that also refines $\mathcal{P}$ and satisfies*

$$\|\mathcal{Q}_0 - \mathcal{Q}\|_1 < \epsilon.$$

Even though we write things for finite alphabets, everything makes good sense for countable partitions as well and the arguments need no adjusting. However, the metric used to compare partitions becomes important since not all metrics on ℓ_1 are equivalent. We use always:

$$\|\mathcal{Q} - \overline{\mathcal{Q}}\|_1 = \sum_j \int_X |1_{Q_j} - 1_{\overline{Q}_j}| d\mu$$

where the partitions $\mathcal{Q}$ and $\overline{\mathcal{Q}}$ are ordered partitions into sets $\{Q_j\}$, $\{\overline{Q}_j\}$ respectively. We also assume that the σ-algebra generated by the partition $\mathcal{P}$ is nonatomic – otherwise there is no real difference between what we did before and what has to be done here.

We shall try to follow the same proof as before. The problem is that when we redefine $\mathcal{Q}_0$ to $\mathcal{Q}$ we are not allowed to change the $\mathcal{P}$-part of the name of points.

That greatly restricts us in the kind of names we are allowed to copy on columns of K-R towers and it is not clear how to proceed. The way to overcome the difficulty is to build the K-R towers inside the uniform algebra generated by $\mathcal{P}$. This being done we look, for example, at our first tower and the first change we wish to make in $\mathcal{Q}_0$. We divide the fibers into a finite number of columns according to the height and according to the $\mathcal{P}$-name.

Next each of these is divided into subcolumns according to the $\mathcal{Q}$-names of points. If a $\mathcal{P}$-column has some good (i.e. good 1-block distribution of $\mathcal{Q}$-names) $\mathcal{Q}$-subcolumn it can be copied onto all the ones that are not good. Next notice that a $\mathcal{P}$-column that contains not even one good $\mathcal{Q}$-name is a set defined in the uniform algebra. Therefore if these sets have small measure then for some large enough N, uniformly over the whole space, we will not encounter these bad columns too many times. It is this N that gives us the uniformity — over all points in the space.

In brief the solution is to change the nature of the uniformity. We do not make all of the columns of the K-R tower good – but we make sure that the bad ones are seen infrequently, uniformly over the whole space. With this remark the proof of the proposition is easily accomplished using the same nested K-R towers as before – **but inside the uniform algebra**.

Finally the J-K theorem is established by constructing a refining sequence of uniform partitions and looking at the inverse limit of the corresponding topological spaces. Notice that if $\mathcal{Q}$ refines $\mathcal{P}$, and both are uniform, then there is a natural homomorphism from $X_\mathcal{Q}$ onto $X_\mathcal{P}$. The way in which the theorem is established also yields a proof of the relative J-K theorem:

THEOREM 3.14. *If* $\mathcal{Y} = (Y, \mathcal{C}, \nu, S) \to (X, \mathcal{B}, \mu, T) = \mathcal{X}$ *are ergodic systems and* $(\widehat{X}, \widehat{T})$ *is a uniquely ergodic model for* $\mathcal{X}$ *then there is a uniquely ergodic model* $(\widehat{Y}, \widehat{S})$ *for* $\mathcal{Y}$ *which topologically extends* $(\widehat{X}, \widehat{T})$.

References

1. P. Halmos, *Lectures on Ergodic Theory*, Chelsea Publ. N.Y., 1956.
2. G. Hansel and J.P. Raoult, *Ergodicity, uniformity and unique ergodicity*, Indiana Univ. Math. J. **23**(1974), 221–237.
3. R.I. Jewett, *The prevalence of uniquely ergodic systems*, J. Math. Mech. **19**(1970), 717–729.
4. W. Krieger, *On unique ergodicity*, Proc. Sixth Berkeley Symposium on Math. Stat. and Prob., 1970, pp. 327–346.

5. K. Petersen, *Ergodic Theory*, Cambridge Univ. Press, 1983.
6. D. Rudolph, *Fundamentals of Measurable Dynamics*, Oxford Science Publ. 1990.

CHAPTER 4

Ergodic and Uniquely Ergodic Orbits

In the previous chapter we discussed various properties of invariant measures
and the systems they define. In this one we shall concentrate on individual points
or orbits and explore the same basic dynamical ideas for them. Recall that a point
$x_0 \in X$ is generic for (X, T, μ) with X compact and μ a T-invariant measure if
for all continuous functions f on X we have

$$\lim_{n\to\infty} \frac{1}{n} \sum_0^{n-1} f(T^i x_0) = \int_X f(x) \, d\mu(x).$$

If μ is an ergodic measure we will call x_0 a T-**ergodic point**, or since clearly this
property is true for any other point on the orbit of x, we will say that the orbit is an
ergodic orbit. If the orbit closure of x_0 is uniquely ergodic then we will say that x_0
is a T-uniquely ergodic point. If the T is fixed in the course of the discussion we will
omit the prefix T. The map T need not be invertible for those notions to make sense
and indeed for most of the chapter we will work in $X = A^{\mathbb{N}}$ with $A = \{1, 2, \dots a\}$
and T the shift on X. This makes the writing and visualization easier. Indeed for
invertible transformations it really makes more sense to use two-sided averages –
over $(-n, n)$ rather than over $[0, n)$.

We need some abbreviations to make the writing more concise, so in this chapter
we denote the block $x(j) \, x(j+1) \dots x(k)$ by x_j^k and for $w \in A^n$, $[w]$ will denote
the cylinder set:

$$[w] = \{x \in X : x_1^n = w\}.$$

If μ is a measure on X and $w \in A^n$ we will write $\mu(w)$ for $\mu([w])$. Basic will be the
idea of the **empirical distribution of k-blocks in an n-block** $u \in A^n$. This is

the measure $\lambda_u^{(k)}$ on A^k defined by:

$$\lambda_u^{(k)}(w) = \frac{1}{n-k+1} \, |\{1 \le j \le n-k+1: \, u_j^{j+k-1} = w\}|.$$

Since indicator functions of finite cylinders in X span the continuous functions it is easy to see that $x \in X$ is generic for μ if and only if for all k, and all $w \in A^k$

$$\lim_{n \to \infty} \lambda_{x_1^n}^{(k)}(w) = \mu(w).$$

In words the empirical distribution of the k-blocks in x_1^n tends to the distribution that μ defines on A^k, i.e. the μ-distribution of x_{m+1}^{m+k} for any m.

LEMMA 4.1. *A point* $y \in A^{\mathbb{N}}$ *is a shift-ergodic point if and only if for all* $k \ge 1$ *and all* $\epsilon > 0$ *there is an* $n_0 = n_0(k, \, \epsilon)$ *such that for all* $n \ge n_0(k, \, \epsilon)$ *there is some* $m_0 = m_0(n)$ *and for all* $m \ge m_0$ *we can* $(1-\epsilon)$*-cover* x_1^m *by disjoint* n*-blocks, so that the empirical* k*-block distribution of each one of them falls within the same* ϵ*-ball in the simplex of measures on* A^k.

We are using the phrase "$(1-\epsilon)$-covers" to mean that the total lengths of these n-blocks is at least $(1 - \epsilon) \cdot m$. Any fixed metric on the simplex of measures on A^k can be used. There are several equivalent variants of this lemma, in particular it is not hard to see that if the property holds for a single n_0 then it will hold for **all** sufficiently large n.

PROOF OF LEMMA 4.1. We begin by showing why points y that are shift-ergodic for μ have this property. Applying the ergodic theorem to the indicator functions of each one of the cylinder sets defined by $w \in A^k$ we find, given $\epsilon > 0$, an n_0 such that for all $n \ge n_0$, there is a set B_n in X with measure greater than $1 - \epsilon/10$, and for all $x \in B_n$, we have

$$(4.1) \qquad \left| \frac{1}{n-k} \sum_0^{n-k-1} 1_{[w]}(T^i x) - \mu(w) \right| < (\epsilon/10) \cdot |A|^{-k}, \quad \text{all } w \in A^k.$$

Note that this inequality depends only on x_1^n and thus we can think of B_n as a subset of A^n. Now, the fact that y is shift-ergodic for μ implies that there is some $m_0(n)$, such that for all $m \ge m_0$,

$$\left| \frac{1}{m-n} \sum_0^{m-n-1} 1_{B_n}(T^i y) - \mu(B_n) \right| < \epsilon/10.$$

This m_0 will serve us for the conclusion of the lemma. All that still needs to be done is to go from a set of starting points in $[0, m - n)$ of n-blocks in B_n that $(1 - \epsilon/10)$- cover to disjoint n-blocks. This is done by simply starting at the first n-block that belongs to B_n, then going to the first n-block that belongs to B_n to the right of this block that is disjoint from it, and so on. Clearly we cover in this way all points in B_n that lie in $[0, m - n)$, and the difference between normalizing by $n - m$, and normalizing by n can be absorbed into the ϵ if n/m is taken less than $\epsilon/10$. Summing (4.1) over $w \in A^k$ shows that the metric on the simplex of measures can be the ℓ_1-metric.

For the other direction, suppose that the condition of the lemma is satisfied for a point y. Consider the measures μ_n defined by the formulae

$$\int f \, d\mu_n = \frac{1}{n} \sum_{i=0}^{n-1} f(T^i y).$$

The property formulated in the lemma shows first of all, that the μ_n have a unique cluster point in the w^*-topology, let us call it μ. If μ would be non ergodic, then there would be some finite cylinder $w \in A^k$ such that the limit of the ergodic averages of $1_{[w]}$ would be **non constant**. This would give, for large n, sets where the distribution of the k-blocks in n-blocks would differ by some fixed amount and this, as is easily seen, contradicts the property assumed in the statement of the lemma. This completes the proof of the lemma. $\square$

In the proof of the first part, if y_1^m were to be divided into blocks, all of them large compared to $n_0(k)$, then an easy averaging argument will show that most of these blocks will consist mostly of the good n_0-blocks in B_{n_0}. Thus we can easily establish the following variant of lemma 1 which will prove useful.

LEMMA 4.2. *A point $y \in A^{\mathbb{N}}$ is shift-ergodic if and only if for all $k \geq 1$ and $\epsilon > 0$ there is an $n_0 = n_0(k, \epsilon)$ and an m_0 so that for all $m \geq m_0$ if x_1^m is divided in any way into disjoint subblocks whose lengths are all greater than n_0 then x_1^m is $(1 - \epsilon)$-covered by blocks whose empirical k-block distributions all lie within a single ϵ-ball in the simplex of measures on A^k.*

The characterization of shift-uniquely ergodic points is similar, but slightly easier to formulate:

LEMMA 4.3. *A point $y \in A^{\mathbb{N}}$ is shift-uniquely ergodic if and only if for all $k \geq 1$ and $\epsilon > 0$ there is some n_0 so that the empirical k-block distribution defined by $\{x_\ell^{\ell+n_0-1}\}_{\ell \in N}$ all lie within a single ϵ-ball in the simplex of measures in A^k.*

The proof of this lemma is carried out exactly like the proof of lemma 1. The fact that we now know that **all** the large n_0-blocks have almost the same k-block distribution is what makes it possible to establish the unique ergodicity of the orbit closure of y if the condition is satisfied.

We will also need the semimetric $\bar{d}$ on $A^{\mathbb{N}}$ which is defined by

$$\bar{d}(x, \, y) = \limsup_{n \to \infty} \frac{1}{n} \, |\{1 \leq j \leq n : \, x(j) \neq y(j)\}|.$$

Notice that this is a semimetric – since $\bar{d}(x, \, y) = 0$ does not imply that $x = y$, only that x and y differ merely in a set of zero upper density. This semimetric is natural since if $\bar{d}(x, \, y) = 0$ then x is a shift-ergodic point if and only if y is also a shift-ergodic point. Now we can formulate our main result which characterizes ergodic points in terms of uniquely ergodic points:

THEOREM 4.4. *A point $x \in A^{\mathbb{N}}$ is a shift-ergodic point if and only if there is a sequence y_i of shift-uniquely ergodic points such that*

$$\lim_{i \to \infty} \, \bar{d}(x, \, y_i) = 0.$$

It is easy to see that the property expressed in the theorem implies that x is a shift-ergodic point using the characterization that we have in lemma 4.1 and lemma 4.3. For the converse we will need the following proposition which will guarantee for us that in the course of constructing uniquely ergodic points that are close to the given ergodic point we do not lose the basic properties of x. Notice that given any finite block map $\phi_k : \, A^k \to B$ we can define

$$\Phi : \, A^{\mathbb{N}} \to B^{\mathbb{N}}$$

by means of the formula

(4.2) $$\Phi(y)(m) = \phi_k(y_m^{m+k-1}).$$

It follows at once from the general fact that when π is a homomorphism of one system $(X_1, \, T_1, \, \mu_1)$ to another $(X_1, \, T_2, \, \mu_2)$, then for any $T_1 - \mu_1$ generic point

y, $\pi(y)$ is $T_2 - \mu_2$ generic, that $\Phi(y)$ is an ergodic point if y is one. A **finitary** mapping Φ is a mapping with the property (4.2) except that while for each y and m, the needed k is finite, it is not uniformly bounded. In this case Φ may not be defined everywhere, but we suppose that it is defined on a set of full μ-measure and it is there that this finitary property is assumed to hold. The proposition says that these mappings preserve ergodicity.

PROPOSITION 4.5. *If y is an ergodic point and Φ is a finitary mapping then $\Phi(y)$ is also an ergodic point.*

In the first proof we give of the theorem we will not need the extra flexibility that finitary maps give.

We turn now to the proof of the less obvious direction of theorem 4.4. To begin with we will carry out the argument under the additional assumption that the ergodic measure μ for y is such that all powers of the shift are ergodic. This is the same as assuming that no rational points lie in the point spectrum of T.

By theorem 3.5 from Chapter 3 we know that the sequence of pairs $(x_1,\ 1)$ $(x_2,\ 2), \ldots (x_k,\ k)\ (x_{k+1},\ 1), \ldots$ is also ergodic for some measure $\tilde{\mu}$ defined now on $(A \times \{1,\ 2, \ldots k\})^{\mathbb{N}}$ with respect to the shift there. Furthermore, for this measure all powers of the shift prime to k are ergodic. Thus any finite block coding that we give will have the property that if we carry it out over the whole y orbit – then we will get again a point which is ergodic and without rational point spectrum except for multiples of k. It is this which will enable us to repeat the step we are about to describe on the modified sequence.

Suppose then that $\epsilon > 0$ is given and that y is a fixed ergodic point. We can find an n_1 large enough so that $(1 - \epsilon/10)$ of the n_1-blocks $y_{Ln_1+1}^{(L+1)n_1}$, $L = 1,\ 2, \ldots$ etc. have a 1-block distribution within $1/10$ of the 1-block distribution of μ. Choose one of these and form a new point $y^{(1)}$ by changing all the bad n_1-blocks to this fixed good one. This $y^{(1)}$ satisfies:

(i) $\bar{d}(y,\ y^{(1)}) < \epsilon/10$

(ii) all of its n_1-blocks ending in positions divisible by n_1 have a 1-block distribution within $1/10$ of a fixed distribution.

(iii) $y^{(1)}$ is an ergodic point.

This last point (iii) follows since $y^{(1)}$ is obtained from the pair sequence $(y_m, m(\bmod n_1))$ by a block code and this point is an ergodic point.

Next we find for $y^{(1)}$ an n_2 which is a multiple of n_1 so that but for an $\epsilon/100$ percentage, the n_2 blocks have a 2-block distribution within $1/100$ of the 2-block distribution of the process defined by $y^{(1)}$. The place where this percentage $\epsilon/100$ begins, say beyond M_2, is where we will begin to make the changes in $y^{(1)}$. This will ensure that the limiting point that we will eventually get is within ϵ of y in $\bar{d}$. This is so because for some M_1, determined when we constructed $y^{(1)}$, and all $m \geq M_1$, $y^{(1)}$ and all subsequent points in our construction will differ from y by less than m ϵ-places in $[1, m]$. Beyond this M_2, change all n_2-block that are not good to one fixed good n_2-block. This defines $y^{(2)}$ as a function of $y^{(1)} \times$ finite rotation and thus $y^{(2)}$ is also an ergodic point. Note further that the n_1-blocks of $y^{(2)}$ are the n_1-blocks of $y^{(1)}$. Thus the uniformity on n_1-blocks from the first step is not lost.

Let me reiterate that there are some bad n_2-blocks in $y^{(2)}$ that survive – however these occur only in the initial M_2-segment. Thus they have 0-measure with respect to the $y^{(2)}$-process. Also there will be many n_3-blocks (in fact all but the initial ones) that avoid these. Thus in the final $y^{(\infty)}$ these blocks do not appear in the ω-limit set of $y^{(\infty)}$ where the essential orbit closure of $y^{(\infty)}$ is.

There is a distinction between uniquely ergodic, which means that there is only one invariant measure, and strict ergodicity which means that in addition, the support of the measure is the whole space. Any uniquely ergodic system is strictly ergodic when restricted to the support of the unique invariant measure. We are constructing uniquely ergodic points — not strictly ergodic points.

The basic procedure of going from $y^{(1)}$ to $y^{(2)}$ can be iterated again and again and will produce a $y^{(\infty)}$ which has a uniquely ergodic orbit closure and is within ϵ in $\bar{d}$ from y. This concludes the proof of the theorem in case all powers of the shift are ergodic. For the general case we can proceed as follows: Choose an irrational α so that no multiple of α is in the point spectrum of the system defined by y. Next take a sequence of intervals on the unit circle J_1, $J_2 \cdots$, that nest down to 1. Fixing a point, say 1, we can look at successive returns to J_1 under rotation by α, this divides the integers into a series of blocks with two lengths depending on the size of J_1. We use these as our blocks instead of blocks of the same size as before. For the next step use successive returns to J_2. Since $J_2 \subset J_1$ this is a subset of the previous sequence – this plays the role of having n_2 a multiple of n_1 in the case that we just treated in detail. With these blocks exactly the same proof can be carried out as before. The ergodicity of each point follows from theorem 3.5 in the previous chapter.

The uniquely ergodic processes that we construct in this way are not necessarily isomorphic to the initial process. In fact they are not even, in general, factors of the original process since we use an auxiliary block structure. The proof can be carried out in such a way that all of the intermediate processes, and the final one as well, are factors of the original process. This is more complicated to carry out and we will only give the main idea.

Now we have to impose a block structure on our system internally. For this we find in our measure space a sequence of sets $C_1 \supset C_2 \supset \ldots$ such that each C_i is the base of a $K - R$ tower, each C_i is in the **algebra** generated by the finite cylinder sets and the minimum column heights over C_i go to infinity. These are found much in the same way as in Chapter 3, and we merely have to remark that the bases can be found in the algebra. We are not asserting that these towers are bounded, however, the return time function is finitary and that will suffice. It is at this point that we need Proposition 4.5. Basically we now combine the technique of the preceding chapter that was used to construct uniform partitions with the orbit changes that we have just employed.

The easy direction of the theorem can be strengthened in the following way:

PROPOSITION 4.6. *The ergodic points form a closed set in the $\overline{d}$-metric.*

Since uniquely ergodic points are easily seen to be ergodic this implies that a $\overline{d}$-limit of uniquely ergodic points is an ergodic point. The proof of the proposition follows immediately from our characterization of ergodic points in lemma 1. Note that the space $A^{\mathbb{N}}$ with the $\overline{d}$-metric is far from being separable. In fact if we fix a partition of the unit circle into two nonempty intervals and look at the orbits of 1 say under different rotations R_α, R_β and note when it visits the two intervals we will get sequences in $\{0, 1\}^{\mathbb{N}}$ that remain a fixed distance apart in $\bar{d}$ – for α, β rationally unrelated numbers.

I would like to add a few more technical remarks concerning strict ergodicity. We can require more of a generic point, namely we can demand that it lie in the support of the measure μ which it generates. Such points were studies by H. Furstenberg who called them regular points. We shall call a shift-ergodic point **strictly generic** if it is generic for an ergodic measure μ and belongs to the support of μ. A point will be called **strictly ergodic** if the closure of its orbit is a strictly ergodic system. Another way of saying this is to require that it be both a uniquely ergodic point and a recurrent point. It is natural to ask whether or not our theorem can be strengthened to say that any generic point for an ergodic measure is a limit of $\bar{d}$ of strictly ergodic points. An elaboration of our construction above shows that this is indeed the case. Let us sketch this briefly. We maintain the notation in the proof of theorem 4.4.

The first step which constructed $y^{(1)}$ remains unchanged. In the second step, n_2 and M_2 are found as before, taking care to have M_2 a multiple of n_1. Then we do the following to construct $y^{(2)}$:

(a) leave $y^{(1)}$ in the first M_2-places unchanged.

(b) beyond M_2, all bad n_2-blocks are changed to one fixed good n_2-block.

(c) For some large multiple of M_2, say M_2^2, at all places $j \cdot M_2^2$, $j = 1, 2, \ldots$ change the M_2-block to the **initial** M_2-block of $y^{(1)}$.

This last change (c), now puts the anomalous initial M_2-block into the support of our new point $y^{(2)}$, which has the required uniformity for the distribution of 2-blocks not in the size n_2-but rather is the size M_2^3.

For the continuation of the process, i.e. in forming $y^{(3)}$, we will no longer change blocks of size M_2^3. Note that $y^{(2)}$ is an ergodic point since it is a function of $y^{(1)}$ and a periodic point with period M_2^3.

It should be clear how to continue this process so that the resulting $y^{(\infty)}$ will now be a strictly ergodic point. This establishes:

THEOREM 4.4'. *Any shift-ergodic point $x \in A^{\mathbb{N}}$ is a limit in $\bar{d}$ of strictly ergodic points.*

This type of construction yields examples of strictly ergodic system in $\{0,1\}^{\mathbb{N}}$ where -1 is in the spectrum but the corresponding eigenfunction is not continuous. This is what would happen if we would apply the above theorem to a point like z defined by concatenating in succession blocks of odd length of the form

$$0[10]^{n^n}, \quad n = 1, 2, \cdots.$$

This point is already uniquely ergodic, but is not strictly ergodic, and a strictly ergodic point, close to it in $\bar{d}$ will have the property that when paired with the periodic sequence $\{01\}^{\mathbb{N}}$ a non generic point is obtained!

A very similar argument will show how to construct for any ergodic point x another y such that the pair (x, y) is not even a generic point. This shows that there are no universal multipliers in this class.

One can view the main theorem of this chapter as an example of theorems of the type:

If an object A has an ϵ-version of some property $\mathcal{P}$ then the reason is that A is **very close** to an object A' that has property $\mathcal{P}$ with $\epsilon = 0$. An elementary example of this type of result is: if a bounded sequence of reals $\{a_n\}$ is an ϵ-homomorphism, i.e.

$$|a_{n+m} - a_n - a_m| < \epsilon, \qquad \text{all } n, \, m$$

then $\{a_n\}$ is within ϵ of a homomorphism, i.e. there is a constant c

$$|a_n - nc| \leq \epsilon, \qquad \text{all } n.$$

There are many more results in analysis of this type and what we have here is a dynamical example of this kind of result.

References

1. D. Kazhdan, *On ϵ-representations*, Israel J. of Math. **43**(1982), 315–323.
2. N.J. Kalton and J.W. Roberts, *Uniformly exhastive submeasures and nearly additive set functions*, Trans. Amer. Math. Soc. **278**(1983), 803–816.
3. E. Arthur Robinson, Jr., *On uniform convergence in the Wiener-Wintner theorem*, J. London Math. Soc. (2) **49**(1994), no. 3, 493–501.
4. Peter Walters, *Topological Wiener-Wintner ergodic theorems and a random L^2 ergodic theorem*, Ergodic Theory Dynam. Systems **16**(1996), no. 1, 179–206.

CHAPTER 5

Translation Invariant Graphs and Recurrence

In this chapter we shall be considering graphs G whose vertices are $\mathbb{Z}$, the integers, and which are invariant under the action of $\mathbb{Z}$ on itself by addition. This means that if (i, j) is an edge of G so is the pair $(i+n, \ j+n)$ for all n. Clearly such a graph is determined by a subset $D \subset \mathbb{N}$ that describes the distances $|i - j|$ that one sees as (i, j) varies over all edges of the graph. The graph with the distance set D will be denoted by $G(D)$. Recall that a coloring of a graph is an assignment of distinct colors to the vertices of G in such a way that adjacent vertices (those that are joined by an edge) are assigned different colors. The **chromatic number** of G is the minimal number of colors needed to color the graph. It is evident that the smaller the set D the more likely it is that the chromatic number of $G(D)$ will be finite. Several years ago, P. Erdos asked the following question:

QUESTION.. *Is the chromatic number of $G(D)$ finite for every lacunary set D?*

Recall that $D = \{1 \leq d_1 < d_2 < d_3 < \ldots\}$ is **lacunary** if $\inf_j \ d_{j+1}/d_j > 1$. Y. Katznelson answered this question in the affirmative by translating it into a dynamical question which he then proceeded to answer. This translation is a good example of single orbit dynamics in action. Let $\mathcal{C}$ be a finite set - the set of colors in a finite coloring of $G(D)$, which we assume for the moment exists. Then a coloring can be expressed as a mapping $\gamma : \mathbb{Z} \to \mathcal{C}$ that is D-**admissible** in the sense that whenever $|i - j| \in D$ we have $\gamma(i) \neq \gamma(j)$. We can think of γ as a point in $\mathcal{C}^{\mathbb{Z}}$ and if $T : \mathcal{C}^{\mathbb{Z}} \to \mathcal{C}^{\mathbb{Z}}$ denotes the coordinate shift we observe that $T^n\gamma$ is also admissible for any $n \in \mathbb{Z}$. Indeed this follows immediately from the fact that $G(D)$ is translation invariant. Thus the existence of a single D-admissible γ gives rise to a whole orbit $\{T^n\gamma\}$ of D-admissible colorings. Next we observe that if $\{\gamma_k\}$ is a sequence of D-admissible colorings and $\gamma_k \to \gamma$ in the product topology on $\mathcal{C}^{\mathbb{Z}}$ then γ is also

D-admissible. This is a general fact - true for any graph on $\mathbb{Z}$. Thus if we denote by $Y \subset \mathcal{C}^{\mathbb{Z}}$ the set of **all** D-admissible colorings we see that Y is a **closed**, T-invariant space and thus the existence of a single D-admissible coloring has given us an entire dynamical system (Y, T). Finally, the D-admissibility of all $y \in Y$ can be expressed as the existence of a uniform lower bound for the distances

$$\rho(T^d y, \, y) \geq b$$

where ρ is a metric on Y and b is chosen so that if the 0th coordinates of y_1, y_2 differ then $\rho(y_1, \, y_2) \geq b$. This suggests the following definition:

DEFINITION 5.1. A set $D \subset \mathbb{N}$ is said to be a non-recurrence set for $(Y, \, \rho, \, T)$ if for some constant $b > 0$ and all $y \in Y$, $d \in D$

$$\rho(T^d y, \, y) \geq b.$$

It will be called simply a **non-recurrence set** if there is some dynamical system $(Y, \, \rho, \, T)$ for which it is a non-recurrence set. Our discussion up to this point has established one direction of the following:

LEMMA 5.2. *The graph $G(D)$ has a finite chromatic number if and only if D is a non-recurrence set.*

PROOF. For the remaining implication suppose that D is a non-recurrence set for $(Y, \, \rho, \, T)$ with constant b. Let $U_1, \ldots U_a$ be a covering of Y by sets of diameter less than $b/2$, and let $y_0 \in Y$ be arbitrary. Define an element $\gamma \in \{1, \, 2, \, \ldots \, a\}^{\mathbb{Z}}$ by the condition that for all n

$$T^n y_0 \in U_{\gamma(n)}$$

and it is then clear that $\gamma(n)$ provides an admissable coloring of $G(D)$ showing that $G(D)$ has a finite chromatic number. $\square$

The case when $G(D)$ has an infinite chromatic number has a simple combinatorial meaning. Indeed, this happens if and only if in any partition of $\mathbb{N}$ into finitely many sets one of the sets contains a pair (i, j) with $|i - j| \in D$. This is the simplest type of a phenomenon that we will come back to later on, namely: what kind of patterns can we expect to find in large sets. An example of a set D exhibiting this kind of behavior is $\{1, 4, 9, \ldots n^2, \ldots\}$. Here even a density version is valid, i.e. if a set $A \subset \mathbb{N}$ has positive upper density then it contains a pair (i, j) with $|i - j|$ a perfect square.

Returning to our non-recurrence sets, an important observation is that if

$$D = \bigcup_{j=1}^{J} D_j$$

and for each $1 \leq j \leq J$, $G(D_j)$ has a finite chromatic number so does $G(D)$. Indeed, if $\gamma_j : \mathbb{Z} \to \mathcal{C}^{(j)}$ are admissible colorings for the $G(D_j)$, the product mapping into $\mathcal{C}^{(1)} \times \mathcal{C}^{(2)} \times \ldots \times \mathcal{C}^{(J)}$ provides an admissible coloring for $G(D)$. Since any lacunary set can be divided into finitely many sequences in each of which the infimum of the ratios of successive members is at least 10 it suffices to show that such thin lacunary sequences are non-recurrence sets. To that end we shall construct an $\alpha \in [0, 2\pi)$ such that for all integers k, $|e^{i\alpha d_k} - 1| \geq 1/10$. Then, the dynamical system consisting of $\mathbb{T}^1$ - the unit circle and rotation by α, R_α, will witness the non-recurrence of $\{d_k\}_1^\infty$.

The construction is simply the observation that the set

$$\bigcap_k \{t : |e^{itd_k} - 1| \geq .1\}$$

contains a non-trivial Cantor set. Indeed, assuming inductively that

$$E_m = \bigcap_{k=1}^{m} \{t : |e^{itd_k} - 1| \geq .1\}$$

contains at least 2^{m-1} intervals of length greater than $\frac{2}{d_m}$, the fact that $d_{m+1} \geq 10 \cdot d_m$ shows that in each one of them we will find at least two intervals of length greater than $\frac{2}{d_{m+1}}$ in which $|e^{itd_{m+1}} - 1| \geq .1$. This continues the induction (which is trivially verified for $m = 1$) and establishes our claim. We have established:

THEOREM 5.3. (Y. Katznelson) If D is lacunary then $G(D)$ has a finite chromatic number.

Our next goal is to show that lacunarity is the optimal growth rate. For this we introduce a notion that is stronger than D not being a non-recurrence sequence.

DEFINITION 5.4. A sequence $\{s_j\}$ is said to be a **Poincaré sequence** if for any finite measure preserving system $(X, \mathcal{B}, \mu, T)$ and any $B \in \mathcal{B}$ with positive measure we have

$$\mu(T^{s_j} B \cap B) > 0 \qquad \text{for some } s_j \text{ in the sequence } \{s_j\}.$$

Poincaré observed that $\mathbb{N}$ has this property and used it to prove his famous recurrence theorem which says that in the above notation $\mu - a.e.$ $x \in B$ returns to B infinitely many times. Since any topological system (Y, ρ, T) has finite invariant measures, μ, a Poincaré sequence cannot be a non-recurrence sequence. Indeed for any presumptive constant $b > 0$ which would witness the non-recurrence of $\{s_j\}$ for (Y, ρ, T), there would have to be a set B with diameter less than b and having positive μ-measure which would give rise to a contradiction. Here is a suficient condition for a sequence to be a Poincaré sequence.

LEMMA 5.5. *If for every $\alpha \in (0, 2\pi)$*

$$\lim_{n\to\infty} \frac{1}{n} \sum_{k=1}^{n} e^{i\alpha s_k} = 0$$

then $\{s_k\}_1^\infty$ is a Poincaré sequence.

PROOF. Let $(X, \mathcal{B}, \mu, T)$ be a measure preserving system and let U be the unitary operator defined on $L^2(X, \mathcal{B}, \mu)$ by the action of T, i.e.

$$(Uf)(x) = f(Tx).$$

Let H_0 denote the subspace of invariant functions and for a set of positive measure B, let f_0 be the projection of 1_B on the invariant functions. Since this can also be seen as a conditional expectation with respect to the σ-algebra of invariant sets $f_0 \geq 0$ and is not zero. Now since $1_B - f_0$ is orthogonal to the space of invariant functions its spectral measure with respect to U doesn't have any atoms at $\{0\}$. Thus from the spectral representation we deduce that in L^2-norm

$$\left\| \frac{1}{n} \sum_{1}^{n} U^{s_k}(1_B - f_0) \right\|_{L^2} \longrightarrow 0$$

or

$$\left\| \left(\frac{1}{n} \sum_{1}^{n} U^{s_k} 1_B \right) - f_0 \right\|_{L_2} \longrightarrow 0$$

and integrating against 1_B and using the fact that f_0 is the projection of 1_B we see that

$$\lim_{n\to\infty} \frac{1}{n} \sum_1^n \mu(B \cap T^{-s_k}B) = \|f_0\|^2 > 0$$

which clearly implies that $\{s_k\}$ is a Poincaré sequence. $\square$

The proof we have just given is in fact non-Neumann's original proof for the mean ergodic theorem. He used the fact that $\mathbb{N}$ satisfies the assumptions of the proposition, which is Weyl's famous theorem on the equidistribution of $\{n\alpha\}$. Indeed, Weyl also showed that the squares satisfy the assumption which explains our earlier remarks that G (squares) has an infinite chromatic number.

Returning to our sublacunarity condition we recall a very nice probalistic construction due to Ajtai-Havas-Komlos: If a_n grows subexponentially, i.e. $\frac{1}{n} \log a_n \to 0$ as $n \to \infty$, and $b_n \nearrow \infty$ with

$$\ldots < a_n < a_n + b_n < a_{n+1} < \ldots$$

then for numbers $d_j \in [a_j, \ a_j + b_j]$ chosen independently and uniformly, with probability one the $\{d_j\}$ - sequence will satisfy

$$\frac{1}{n} \sum^{n_1} e^{i\alpha d_j} \longrightarrow 0 \text{ for all } \alpha \in (0, \ 2\pi).$$

Thus for any rate of growth that is less than exponential (which is essentially what lacunarity is all about) there are sequences D of that growth rate with $G(D)$ having an infinite chromatic number, and this shows that from the point of view of growth conditions theorem 5.3 is optimal.

Let us now come back to the topological version and for brevity let us say that D is a **recurrence set** if $G(D)$ has an infinite chromatic number. This means that for any dynamical system $(Y, \ \rho, \ T)$ and any $\epsilon > 0$ there is a point y_0 and a $d \in D$ with

$$\rho(T^d y_0, \ y_0) < \epsilon.$$

Since any system contains minimal sets it suffices to restrict attention here to minimal systems. For minimal systems the set of such y's is a dense open set.

To see this fact, let U be an open set. By the minimality there is some N such that for any $y \in Y$, and some $0 \leq n \leq N$, we have $T^n y \in U$. Using the uniform continuity of T^n, we find now a $\delta > 0$ such that if $\rho(u, v) < \delta$ then for all $0 \le n \le N$

$$\rho(T^n u, T^n v) < \epsilon.$$

Now let z_0 be a point in Y and $d_0 \in D$ such that

$$(*) \qquad \rho(T^{d_0} z_0, z_0) < \delta.$$

For some $0 \le n_0 \le N$ we have $T^{n_0} z_0 = y_0 \in U$ and from $(*)$ we get $\rho(T^{d_0} y_0, y_0) < \epsilon$. Thus points that ϵ return form an open dense set. Intersecting over $\epsilon \to 0$ gives a dense G_δ in Y of points y for which

$$\inf_{d \in D} \rho(T^d y, y) = 0.$$

Thus there are points which actually recur along a given recurrence set.

A natural question that arises from the proof of the theorem above is:

PROBLEM. *If D is a recurrence sequence for all group rotations is it a recurrence set?*

A little bit of evidence for a positive answer to that problem comes from looking at a slightly different characterization of recurrence sets. Let $\mathcal{N}$ denote the collection of sets of the form

$$N(U, U) = \{n : T^{-n} U \cap U \ne \emptyset, \ U \text{ open}\}$$

where T is a minimal transformation. Denote by $\mathcal{N}^*$ the subsets of $\mathbb{N}$ that have a non-empty intersection with every element of $\mathcal{N}$. Then $\mathcal{N}^*$ is exactly the class of recurrence sets. For minimal transformations, another description of $N(U, U)$ is obtained by fixing some y_0 and denoting

$$N(y_0, U) = \{n : T^n y_0 \in U\}$$

Then $N(U, U) = N(y_0, U) - N(y_0, U)$. Notice that the minimality of T implies that $N(y_0, U)$ is a **syndetic** set (a set with bounded gaps) and so any $N(U, U)$ is the set of differences of a syndetic set. Conversely, if S is a syndetic set and ω denotes its indicator function in $\{0, 1\}^{\mathbb{Z}}$ then if Y is any minimal set in the orbit closure of ω, and $U = \{y : y(0) = 1\}$, the fact that S is syndetic implies that U

is non empty and $N(U, U) \subset S - S$. Thus $\mathcal{N}$ consists essentially of all sets of the form $S - S$ where S is a syndetic set.

Recall that the Bohr group is the compactification of $\mathbb{Z}$ obtained by looking at the dual group of $\mathbb{T}^1$ viewed as a discrete abelian group. W. Veech proved in [V-1968] that any set of the form $S - S$ with S syndetic contains a neighborhood of the identity in the Bohr compactification **up to a set of zero density**. It is not known if that zero density set can be omitted. If it could then a positive answer to the above problem would follow.

We conclude this chapter with two more remarks concerning recurrence.

REMARK 1. Connections with thin sets in Harmonic analysis.

Lacunary sets are a basic example of what have come to be known as Sidon sets of $\mathbb{Z}$.

Recall that a subset $E \subset \mathbb{Z}$ is a Sidon set if there is a constant C and any trigonometric polynomial P with exponents belonging to E satisfies

$$\sum_n |\hat{P}(n)| \leq C\|P\|_\infty.$$

A natural guess is that any Sidon set is a set of non-recurrence. This seems to be related to some of the difficult open problems concerning the nature of Sidon sets. Here is a piece of evidence in favor of a positive answer. Our entire discussion of graphs and their chromatic number makes perfectly good sense for any discrete abelian group Γ. If D is a symmetric subset of Γ then let $G(D)$ denote the graph whose vertex set is Γ and with edges - all pairs (γ_1, γ_2) such that $\gamma_1 - \gamma_2 \in D$. If $\Gamma = \sum_1^\infty \mathbb{Z}/p\mathbb{Z}$, i.e. a countable sum of cyclic groups of order p, a prime, then it is a theorem of M.-P. Malliavin - Bramere and P. Malliavin [MM-1967] that any Sidon set in Γ is a finite union of independent sets. Since independent sets are clearly sets of non-recurrence (for Γ-actions now) for these groups it is the case that Sidon sets are sets of non-recurrence.

This question is related to another old problem in the theory of Sidon sets. Namely, is there a Sidon set which is also dense in the Bohr compactification. A recent result concerning this question is due to T. Ramsey who showed that if there is a Sidon set in $\mathbb{Z}$ that clusters at 0 in the Bohr compactification then there is a Sidon set that is dense in the Bohr compactification.

The evidence of random set constructions [K-1973] is that Sidon sets are necessarily nowhere dense in the Bohr compactification, and this is further evidence that Sidon sets in $\mathbb{Z}$ are indeed sets of non-recurrence. Finally it should be remarked that in a trivial fashion the odd integers $2\mathbb{Z} + 1$ are an example of a non-recurrence set that is certainly far too large to be a Sidon set.

REMARK 2. Poincaré sets and sets of recurrence

We have seen that a Poincaré set is necessarily a set of recurrence. The converse is not true in general. The significance of this will become clearer when we discuss more general results concerning patterns in large sets in the next chapter. For now I would like to give a brief description of the construction by I. Kriz [Kr-1987] of a set that is a set of recurrence but not a Poincaré set. At the heart of Kriz's example lie the Kneser graphs $K(n, d)$. These are graphs with vertex set $V(n, d)$ equal to the $\binom{2n+d}{n}$ subsets of $\{1, \ 2, \ \ldots \ 2n + d\}$ of size n and edge set defined by putting an edge between $A, \ B \ \in V(n, \ d)$ if and only if $A \cap B = \emptyset$.

The following coloring:

$$C_j = \{A : \ \min A = j\} \qquad 1 \le j \le d + 1$$

$$C_{d+2} = \{A : \ \min A \ge d + 2\}$$

is clearly admissible and thus the chromatic numer of $K(n, \ d)$ is at most $d + 2$. L. Lovasz [Lo-1978] showed that the chromatic number of $K(n, \ d)$ equals $d + 2$. In spite of the elementary combinatorial nature of this statement the proof by Lovasz, as well as a later proof by I. Barany depend in an essential way on a non-trivial topological property of S^d - one of the antipodal point theorems of K. Borsuk!

Here is how one goes about imbedding this graph in the integers:

Take very large primes p_1, p_2, ... p_{2n+d} and set $M = p_1 p_2 \ldots p_{2n+d}$. Choose N large compared to M and set

$$A = \{a \in [M, (N-1)M] : a \equiv \text{ even integer } (\neq 0)(\mathrm{mod} p_i)$$

$$\text{for fewer than } n \text{ indices } i,$$

$$a \equiv \text{ odd integer } (\mathrm{not} \pm 1)(\mathrm{mod} p_i) \text{ for the rest}\}$$

$$B = \{b \in [M, (N-1) M] : b \equiv \text{ odd integer}(\mathrm{mod} p_i) \text{ for fewer than } n \text{ indices } i$$

$$b \equiv \text{ even integer } (\neq 0)(\mathrm{mod} p_i) \text{ for the rest}\}$$

and let

$$C = \{c \in [0, M] : c \equiv 1 \text{ or } (p_i - 1)(\mathrm{mod} \ p_i) \text{ for more than } 2n \text{ indices } i\}.$$

With a proper choice of the parameters, both A and B have density greater than $\frac{1}{2} - \epsilon$ and together fill up most $[0, MN]$. Here d is fixed, then n is taken very large, then the $p_1, \ldots p_{2n+d}$ are much larger etc. Furthermore one sees that both A and B are C-independent, i.e. $\begin{array}{l} (A + C) \cap A = \emptyset \\ (B + C) \cap B = \emptyset \end{array}$.

Finally letting

$$V = \{v \in [0, M] : v \equiv 2(\mathrm{mod} \ p_i) \text{ for exactly } n \text{ indices } i,$$

$$v \equiv 1(\mathrm{mod} \ p_i) \text{ for the others}\}$$

we see that V corresponds to the vertices of $K(n, d)$ and the edges are given by elements of C so that the chromatic number of $G(C) \geq d + 2$.

At this point we see the main idea: d can be arbitrarily large, and we are constructing a set C so that the chromatic number of $G(C)$ is greater than d, but there are quite large sets - A and B each of density almost $1/2$ that together cover most of $[0, MN]$ and such that each of them is C-independent. What remains is to put together an infinite sequence of such examples in such a way as to get a C with $G(C)$ having an infinite chromatic number - so that C is a recurrence set, but not a Poincaré set since there will also be a positive density set which is C-independent.

References

1. M. Ajtai, I. Havas and J. Komlós, *Every group admits a bad topology*, Studies in pure mathematics, 21–34, Birkhäuser, Basel-Boston, Mass., 1983.
2. I. Bárány, *A short proof of Kneser's conjecture*, J. Combin. Theory Ser. A **25**(1978), no. 3, 325–326.
3. J. Bourgain, *Rusza's problem on sets of recurrence*, Israel J. of Math. **59**(1987), 150–166.
4. Y. Katznelson, *Sequences of Integers Dense in the Bohr Group* (June 1973), Proc. Royal Inst. of Technology (Sweden), 73–86.
5. I. Kriz, *Large independent sets in shift-invariant graphs. Solution of Bergelson's problem*, Graphs and Combinatorics **3** (1987), 145–158.
6. L. Lovász, *Kneser's conjecture, chromatic number and homotopy*, J. Comb. Theory (A) **25**(1978), 319–324.
7. T. Ramsey, *Bohr cluster points of Sidon sets*, Coll. Math. **48**(1995), 285–290.
8. R. McCutcheon, *Three results in recurrence. Ergodic theory and its connections with harmonic analysis* (Alexandria, 1993), 349–358, London Math. Soc. Lecture Note Ser., **205**, Cambridge, 1995.
9. Marie-Paule Malliavin-Brameret and Paul Malliavin, *Caracterisation arithmetique d'une classe d'ensembles de Helson*, C.R. Acad. Sci. Paris Ser. A-B **264**(1967) A192–A193.
10. William A. Veech, *The equicontinuous structure relation for minimal Abelian transformation groups*, Amer. J. Math. **90**(1968), 723–732.

CHAPTER 6

Patterns in Large Sets

Scientists in the natural sciences as well as scholars in the humanities deal with patterns. Many discoveries take the form of the unveilings of new patterns that had not been noticed before. It is important to know when these patterns represent some genuine underlying structure and when they are but artifacts of the type that one would inevitably see even in quite arbitrary data. There is a collection of results that have the form "a sufficiently large set necessarily contains patterns of a special nature". In the previous chapter we encountered a result of that type. Indeed if we understand by a "sufficiently large set of integers" a **syndetic set**, that is one with bounded gaps, and if D is a recurrence set, then the fact that $G(D)$ has an infinite chromatic number implies that syndetic set A necessarily contains a pair of elements $\{a, b\}$ with $|a-b|$ an element of D. This collection $\{\ \{a,\ b\} :\ |a-b| \in D\}$ may be taken to be an example of a pattern of a special nature.

The implication goes as follows. The fact that A is syndetic means exactly that for some J the sets

$$A,\ A+1,\ A+2,\ \ldots\ A+(J-1)$$

cover $\mathbb{Z}$. Let's define a J-coloring of $\mathbb{Z}$ by assigning to $n \in \mathbb{Z}$ the least $0 \le j < J$ such that $n \in A+j$. Since $G(D)$ has an infinite chromatic number, for one of the colors, say j_0, we must have a pair of vertices $\{a,\ b\}$ with $|a-b| \in D$ and both a and b have the color j_0. This means that $a,\ b \in A+j_0$ and then $a' = a - j_0,\ b' = b - j_0$ both belong to A and satisfy $|a' - b'| \in D$ as required.

It is not hard to see that for syndetic sets any collection of finite special patterns that is translation invariant has a dynamical analogue so that the occurrence of these special patterns in syndetic sets is equivalent to some dynamical property of all compact systems $(X,\ T)$. A well-known example of this is the collection of finite arithmetical progressions and the theorem of van der Waerden to the effect that not only do syndetic sets necessarily contain arithmetic progressions of arbitrary length but in any finite partition of the integers one of the sets contains arithmetic progressions of arbitrary length. In fact this apparently stronger statement is already implied by the one concerning syndetic sets. Since this is a nice illustration of the concept of minimal sets we digress briefly to prove this before continuing our discussion with another type of large sets - the subsets of $\mathbb{N}$ with positive density.

PROPOSITION 6.1. *Let $\mathcal{P}$ be a translation invariant collection of finite subsets of $\mathbb{Z}$. If every syndetic set A necessarily contains an element of $\mathcal{P}$ then also in any finite covering $\{C_1, C_2, \ldots C_J\}$ of $\mathbb{Z}$ one of the sets C_{j_0} contains an element of $\mathcal{P}$.*

PROOF. Define an element $\omega_0 \in \{1, 2, \ldots J\}^{\mathbb{Z}}$ by setting $\omega_0(n)$ to be the least j such that $n \in C_j$. Let X denote the orbit closure of ω_0 under the shift T. The system (X, T) has a **minimal** subsystem (X_0, T) and we let $\{x_0(n)\}$ be any element of X_0. If $x_0(0) = j_0$ then the set $\{n : x_0(n) = j_0\} = A$ is syndetic and thus A contains some element of $\mathcal{P}$, say P. Since P is finite it occurs in x_0 in a finite interval say from $-N$ to N. But x_0 being in the orbit closure of ω_0, for some M,

$$\{\omega_0(M + n) : |n| \leq N\} = \{x_0(n) : |n| \leq N\}$$

and thus a translate of P occurs in ω_0 at places in C_{j_0}, i.e. a translate of P is in C_{j_0}. Since $\mathcal{P}$ was translation invariant we are done. $\square$

We turn now to density questions. Let us say that $A \subset \mathbb{Z}$ has **positive upper density** (or simply is **substantial**) if for some $k_i \nearrow \infty$ and $n_i \in \mathbb{Z}$

$$\lim_{i \to \infty} \frac{1}{k_i} \sum_{j=n_i+1}^{n_i+k_i} 1_A(j) = b > 0.$$

It turns out that the dynamical property of a set D that we called being a Poincaré set is equivalent to the assertion that any substantial set A contains a pair $\{u, v\}$ with $|u - v| \in D$. Note that syndetic sets are certainly substantial and so the fact that Poincaré sets are necessarily recurrence sets is immediate in this formulation. The equivalence above is an example of what Furstenberg called the correspondence principle and plays a fundamental role in the application of ergodic theoretic methods to combinatorial number theory. Let us describe this more formally in terms of a translation invariant collection, $\mathcal{P}$ of finite subsets of $\mathbb{Z}$.

PROPOSITION 6.2. *Sets in $\mathbb{Z}$ with positive upper density necessarily contain an element of $\mathcal{P}$, if and only if, for any probability preserving system $(X, \mathcal{B}, \mu, T)$ and any subset $B \in \mathcal{B}$ with positive measure there is an element $P \in \mathcal{P}$, $P = \{u_1, u_2, \ldots u_k\}$ with*

$$(*) \qquad\qquad \mu \left(\bigcap_{i=1}^{k} T^{-u_i} B \right) > 0.$$

PROOF.

1. Assume first the statement for sets in $\mathbb{Z}$. A standard application of the ergodic decomposition shows that it is enough to prove the dynamical statement for ergodic systems. For such systems, for $\mu - a.e.$ x we have that the set $\{n \geq 0 : T^n x \in B\}$ has positive density. Thus for $\mu - a.e.$ x there is an element $P(x) \in \mathcal{P}$ such that for all n in $P(x)$, $T^n x \in B$. Since there is only a countable number of possibilities for P there must be a set of positive measure of x's with the same P_0, and for this P_0 $(*)$ is satisfied. We remark in passing that if there is a bound on the size of the elements in $\mathcal{P}$ then one can easily avoid the use of the ergodic theorem and prove things for finitely additive measures as well.

2. Next we assume the dynamical statement. Given a substantial set $A \subset \mathbb{Z}$ we construct a measure μ on $\{0, 1\}^{\mathbb{Z}}$ as a cluster point in the space of probability measures on $\{0, 1\}^{\mathbb{Z}}$ of the discrete measures

$$\frac{1}{k_j} \sum_{i=n_j+1}^{n_j+k_j} \delta_{T^i x_0}$$

with $x_0 = 1_A$, and the k_j's, n_j's satisfying the defining property of A being a substantial set. This assumption gives the fact that $\mu([1]) > 0$ where $[1]$ is the cylinder set of points with 1 in the 0-th coordinate. The fact that the $k_j \nearrow \infty$ gives easily that μ is shift invariant and it is then an easy matter to see that the dynamical statement gives the desired result. $\qquad\square$

This proposition shows how the whole theory of measure preserving systems can be brought to bear in this "single set" or single orbit question concerning patterns in A. We saw an elementary example of this in the previous chapter when we used the spectral theory of unitary operators to translate properties of equidistribution on the circle to mean ergodic theorems and then to the statement that $\{n^2\}$ is a Poincaré set and thus a recurrence set. Using much deeper structural information about measure preserving systems Furstenberg was able to use this principle to give a dynamical proof that substantial sets in $\mathbb{Z}$ necessarily contain arithmetical progressions of arbitrary length. Later on we were able to extend this to some patterns of polynomial type and recently V. Bergelson and A. Leibman have obtained very general results for polynomial type patterns. We cannot pursue this much more in the present framework and I will devote the rest of the chapter to a discussion of these kinds of questions for $\mathbb{R}^n$ where new kinds of issues come up.

For measurable subsets of $\mathbb{R}$, A, we will say that they have **positive upper density** if for some $k_i \nearrow \infty$, n_i we have

$$\lim_{i \to \infty} \frac{1}{k_i} \int_{n_i}^{n_i+k_i} 1_A(t)dt = b > 0.$$

Measure preserving $\mathbb{R}$ actions on $(X, \mathcal{B}, \mu)$ are defined to be measurable mappings from $\mathbb{R} \times X$ to X given by

$$(t, x) \mapsto (T_t\, x)$$

such that each T_t preserves μ and, of course, $T_t\, T_s = T_{t+s}$. In order to develop a reasonable theory we must assume that $T_t\, x$ is jointly measurable in t and x. This measurability implies a weak sort of continuity for T_t which in turn means that the sets of $\mathbb{R}$ that one sees as visit times to sets of positive measure, $\{t : T_t\, x \in B\} = V$, are a rather special kind of sets of positive upper density. There is some kind of uniformity in the nature of the measurability of V at ∞ which is simply not present in arbitrary subsets of positive density. This means that dynamical theorems for $\mathbb{R}$ actions do not, automatically, give rise to theorems about sets of positive upper density.

I shall illustrate this by a pair of theorems about subsets of positive density in the plane, $\mathbb{R}^2$. The definition for $\mathbb{R}^2$ of positive density is just as for $\mathbb{R}$ but one replaces large intervals $[n_i, \ n_i + k_i]$ by large squares - or large disks.

THEOREM 6.3. *If $E \subset \mathbb{R}^2$ has positive upper density there exists an ℓ_0 such that for all $\ell > \ell_0$ there are points $x, \ y \in E$ with $\|x - y\| = \ell$.*

The norm $\| \cdot \|$ in the theorem is the Euclidean norm, although any strictly convex norm, such as $\ell_p, \ 1 < p < \infty$ will do as well. We tried to prove an analogous theorem for triangles but succeeded only in showing:

THEOREM 6.4. *Let $E \subset \mathbb{R}^2$ have positive upper density, let $\delta > \infty$, and set*

$$E_\delta = \{x \in \mathbb{R}^2 : \ \|x - y\| < \delta \ for \ some \ y \in E\}.$$

Then for any $u, \ v \in \mathbb{R}^2$ there exists an ℓ_0, and for all $\ell > \ell_0$ there is a triple $\{x, \ y, \ z\} \in E_\delta$ such that the triangle with vertices $\{x, \ y, \ z\}$, $\triangle(x, \ y, \ z)$ is congruent to $\triangle(0, \ \ell u, \ \ell v)$.

Indeed it is not possible in this theorem to replace E_δ by E as the following example, due to J. Bourgain, shows. Let $v = 2u$, so we are looking for degenerate isosceles triangles. Let

$$E = \{ \ (x_1, \ x_2) : \ |x_1^2 + x_2^2 - n| \ < \frac{1}{10} \ for \ some \ integer \ n \ \}.$$

The set E consists of circular shells that fill up $1/5$ of the plane, but get thinner and thinner as you move out to infinity. Thus E lack the kind of uniform measurability that one sees for sets coming from visit times of measurable actions. Indeed one does not have in E itself **all** sufficiently large degenerate isosceles triangles as the following reasoning shows. If $x, \ y \in E$ the parallelogram law gives

$$\|x\|^2 + \|y\|^2 - 2\|\frac{x + y}{2}\|^2 = 2\|\frac{x - y}{2}\|^2$$

so that if also $\frac{x+y}{2} \in E$ then the left-hand side is close to an integer. In particular $2\|\frac{x-y}{2}\|^2$ cannot be close to half-integer values so that the difference $\|x - y\|$ cannot take on all large values.

I will give now an overview of the proofs of these two theorems, for full details see [FKW - 1990]. A little more care is needed to set up the basic machinery since the space $\{0, 1\}^{\mathbb{R}^2}$ is far too large. We write things for $\mathbb{R}^2$, but all of this works for any locally compact group. As above, $\| \cdot \|$ will denote the Euclidean metric on $\mathbb{R}^2$. If $E \subset \mathbb{R}^2$ we set

$$\phi(u) = \min(\operatorname{dist}(u, \ E), \ 1)$$

where

$$\operatorname{dist}(u, \ E) = \inf\{\|u - v\| : \ v \in E\}.$$

The family $\{\phi_v(u) = \phi(u + v)\}$ forms an equicontinuous family since ϕ is bounded and Lipschitz with constant equal to one, and thus if X denotes the closure of this family in the topology of uniform convergence on bounded sets X is a compact space. Note that $\psi \in X$ means that there is a sequence of translates of $\phi_1 \ \phi_{v_n} \ (u)$ that tend to $\psi(u)$ uniformly on any disk $\{u : \ \|u\| \le R\}$.

Since X is a function space on $\mathbb{R}^2$ which is translation invariant there is a natural action of $\mathbb{R}^2$ on this space $(T_v \psi)(u) = \psi(u + v)$. Our original ϕ belongs to X and its $\mathbb{R}^2$ orbit is dense. If E is measurable with positive upper density, there is a sequence of squares $S_n \subset \mathbb{R}^2$ with side lengths ℓ_n going to infinity such that

$$m(S_n \cap E)/m \ (S_n) \to b > 0$$

where m denotes Lebesgue measure. Define a measure μ_n on X by the formula

$$\int f(\psi) d\mu_n(\psi) = \frac{1}{m(S_n)} \int\limits_{S_n} f(T_v \phi) dm(v)$$

where f is an arbitrary continuous function on X. Since X is compact metric a subsequence $\{\mu_{n_k}\}$ will converge in the w^* topology to a measure μ. The fact that the side lengths, ℓ_n, of the squares goes to infinity implies that the limiting measure μ is invariant under the $\mathbb{R}^2$-action since for each n and fixed translation $t_0 \in \mathbb{R}^2$ one sees that

$$\left| \int\limits_X f \ d\mu_n - \int\limits_X f \cdot T_{t_0} \ d\mu_n \right| \le \frac{4 \cdot \|t_0\|}{\ell_n} \cdot \|f\|_\infty.$$

Define a subset $\widehat{E} \subset X$ by

$$\widehat{E} = \{\psi \in X : \ \psi(0) = 0\}.$$

Since the function $f_0(\psi) = \psi(0)$ is continuous on X, $\widehat{E}$ is a closed set and

$(*)$
$$\mu(\widehat{E}) = \lim_{j \to \infty} \int_X (1 - f_0)^j \, d\mu.$$

LEMMA 6.5. $\mu(\widehat{E}) \geq b$

PROOF. By $(*)$ it suffices to show that for any j

$$\int_X (1 - f_0)^j \, d\mu \geq b.$$

According to the definition of μ we have

$$\int_X (1 - f_0)^j \, d\mu = \lim_{k \to \infty} \frac{1}{m(S_{n_k})} \int_{S_{n_k}} (1 - f_0(T_v \phi))^j \, dm(v)$$

$$= \lim_{k \to \infty} \frac{1}{m(S_{n_k})} \int_{S_{n_k}} (1 - \phi(v))^j \, dm(v).$$

For $u \in E$, $\phi(u) = 0$ and therefore the last integral is at least

$$\lim_{k \to \infty} \frac{1}{m(S_{n_k})} m(S_{n_k} \cap E) = b.$$

$\square$

We will now see what type of information we can get about patterns in E from the positive μ measure of intersections of sets in X.

PROPOSITION 6.6. If for the elements of $\mathbb{R}^2$, $\{u_1, u_2, \ldots u_k\}$ we have $\mu(\widehat{E} \cap T_{u_1}^{-1} \widehat{E} \cap T_{u_2}^{-1} \widehat{E} \cap \ldots \cap T_{u_k}^{-1} \widehat{E}) > 0$ then for all $\delta > 0$,

$$E_\delta \cap E_\delta - u_1 \cap E_\delta - u_2 \cap \ldots \cap E_\delta - u_k \neq \emptyset$$

i.e. there is some $e_0 \in E_\delta$ such that for all $1 \leq j \leq k$ $\quad e_0 + u_j \in E_\delta$.

PROOF. Define $g(\psi)$ on X by

$$g(\psi) = \begin{cases} \delta - f_0(\psi) & \text{if } f_0(\psi) \leq \delta \\ 0 & \text{if } f_0(\psi) \geq \delta \end{cases}$$

For $\psi \in \widehat{E}$, $g(\psi) > 0$ and thus our hypothesis implies that

$$\int g(\psi) \, g(T_{u_1} \psi) \, g(T_{u_2} \psi) \cdots g(T_{u_k} \psi) \, d\mu(\psi) > 0.$$

From the definition of μ we deduce that for some ℓ,

$$\int_{S_{n_\ell}} g(T_v \, \phi) \, g(T_{u_1+v}\phi) \, \cdots \, g(T_{u_k+v} \, \phi) \, dv > 0.$$

Clearly the last integrand has positive values and $g(T_u \, \phi) > 0$ if and only if $\phi(u) < \delta$ which means exactly that $u \in E_\delta$. This gives us values of u for which simultaneously $u \in E$ and $u + u_j \in E$ for all $1 \leq j \leq k$ as required. $\qquad\square$

This proposition explains what type of result one gets for subsets of the plane from corresponding results on measure preserving actions of $\mathbb{R}^2$. The lack of the perfect correspondence familiar to us from the action of discrete groups is due to the necessity of dealing with continuous functions in the selection theorems. In any event it is now clear that theorem 6.4 can be the consequence of a theorem on dynamical systems. Before stating that we wish to discuss theorem 6.3. Its proof will involve a dynamical theorem which is rather straightforward, and then a more detailed analysis of the structure of subsets of $\mathbb{R}^2$ with positive upper density. This analysis is of a geometric nature and has something of the flavor of the renormalization techniques that have proved to be very useful. This is really not in the main line of our development so I will not discuss that any more here. The dynamical theorem needed for proving the δ-version of theorem 6.3 is the following:

In the formulation we identify $\mathbb{R}^2$ with $\mathbb{C}$ and use complex notation.

THEOREM 6.7. *Let T_z be an $\mathbb{R}^2$ action on a probability space $(X, \mathcal{B}, \mu)$ that preserves the measure μ. Then if P denotes the projection of $L^2(X, \mathcal{B}, \mu)$ onto the $\{T_z\}$ invariant functions we have for any $f \in L^2$*

$$\lim_{R \to \infty} \|\frac{1}{\beta - \alpha} \int_\alpha^\beta T_{Re^{i\theta}} f \, d\theta - Pf\|_{L^2} = 0$$

for any $\alpha < \beta$.

PROOF. Using the spectral representation of the unitary group induced by the $\mathbb{R}^2$ action on $L^2(X, \mathcal{B}, \mu)$ one sees that it suffices to prove the following:

For any $(\xi, \eta) \neq 0$

$$\lim_{R \to \infty} \frac{1}{\beta - \alpha} \int_\eta^\beta e^{i \, [(R\cos\theta)\xi + (R\sin\theta)\eta]} d\theta = 0.$$

This follows immediately from van der Corput's lemma [Zy-1959] and the proof is done.   □

The dynamical theorem behind theorem 6.4 is the following:

THEOREM 6.8. *Let T_z be an $\mathbb{R}^2$ action on a probability space $(X, \mathcal{B}, \mu)$. Let $A \in \mathcal{B}$ be a set with positive measure and $w \in \mathbb{C}$. Then there exists an $\ell_0 > 0$ and for all $\ell \geq \ell_0$ there is a $z \in \mathbb{C}$ with $|z| = \ell$ satisfying*

$$\mu(A \cap T_z A \cap T_{wz} A) > 0.$$

For more background and references we refer the reader to our paper with H. Furstenberg and Y. Katznelson.

References

1. V. Bergelson and A. Leibman, *Polynomial extensions of van der Waerden's and Szemerédi's theorems*, J. Amer. Math. Soc. **9**(1996), o. 3, 725–753.
2. J. Bourgain, *A Szemeredi type theorem for sets of positive density in $\mathbb{R}^k$*, Israel J. of Math **54**(1986), 307–316.
3. H. Furstenberg and B. Weiss, *A Mean Ergodic Theorem for $\frac{1}{N} \sum_{n=1}^{N} (T^n x) g(T^{n^2} x)$*, Convergence in Ergodic Theory and Probability, W. de Gruyten, 1996, 193–227.
4. H. Furstenberg, Y. Katznelson and B. Weiss, *Ergodic theory and configurations in sets of positive density*, in Mathematics of Ramsey Theory, ed. by J. Nesetril and V. Rodl, Springer, 1990, pp. 184–199.
5. A. Zygmund, *Trigonometric series. 2nd ed. Vols. I, II*, Cambridge University Press, New York, 1959.

CHAPTER 7

Entropy and Disjointness

Here is a rapid review of the basic definitions and properties of the Shannon entropy of stochastic processes. We shall freely interchange and switch between the language of random variables x_i, y_j defined on a probability space $(X, \mathcal{B}, \mu)$ and the partitions $\mathcal{P}_i$, $\mathcal{Q}_j$ of X defined by the sets of constancy of these random variables. To the pair of random variables (x, y) corresponds the **join** or **span** of the partitions $\mathcal{P}$, $\mathcal{Q}$ written $\mathcal{P} \vee \mathcal{Q}$ whose atoms are the intersection $P \cap Q$ as P, Q range over the atoms of the partitions $\mathcal{P}$ and $\mathcal{Q}$. The basic definition is for the information content or **entropy** of a random variable x or the partition $\mathcal{P}$ it defines as:

$$H(x) = H(\mathcal{P}) = -\sum p_k \, \log p_k$$

where the p_k are the probabilities attached to the different possible outcomes of x. We are of course considering only discrete valued random variables which for the most part will take on only finitely many values. The quantity $H(x)$ should be thought of as the amount of information that an observer obtains when he sees the outcome of an experiment x. Some basic properties which follow easily from the convexity of $-t \, \log t$ are as follows:

(1) $H(x) \leq \log N$ if at most N of p_k's differ from zero

with equality only in case $p_1 = p_2 = \ldots p_N = \dfrac{1}{N}$

(2) $$H(x, \, y) \leq H(x) + H(y)$$

with equality if and only if x and y are independent.

The conditional information of x with respect to y is defined by

$$H(x|y) = \sum_\eta P(y = \eta) \left\{ -\sum_\xi P(x = \xi|y = \eta) \, \log P(x = \xi|y = \eta) \right\}$$

where the sums are over the possible values of x and y respectively, and $P(x = \xi|y = \eta)$ denotes the conditional probability. For the conditional entropy we have:

$$(3) \qquad\qquad\qquad H(x, \, y) = H(x|y) + H(y)$$

$$(4) \qquad\qquad\qquad H(x|y) \leq H(x)$$

with equality if and only if x and y are independent.

For a stationary process $\{x_j\}_1^\infty$ the subadditivity of (2) gives easily that

$$H(x_1, \, x_2, \, \ldots \, x_{n+m}) \leq H(x_1, \, \ldots \, x_n) + H(x_1, \, \ldots \, x_m)$$

and then a well known calculus exercise shows that

$$H(\{x_j\}_1^\infty) = \lim_{n\to\infty} \frac{1}{n} H(x_1, \, \ldots \, x_n)$$

is well defined. This is what Shannon called the average information contained in the process $\{x_j\}_1^\infty$. It is now called simply the **entropy** of the **process**. One can extend (3) by induction to yield the following:

$$H(x_1, \, x_2, \, \ldots \, x_n) = H(x_1|x_2, \, \ldots \, x_n) + H(x_2|x_3, \, \ldots \, x_n) \, + \ldots$$
$$(5) \qquad\qquad\qquad + H(x_{n-1}|x_n) + H(x_n).$$

Using the stationarity one gets the following alternative expression for the entropy of a process:

$$H(\{x_j\}_1^\infty) = \lim_{n\to\infty} H(x_1|x_2, \, x_3, \, \ldots \, x_n).$$

It is not hard to give a meaning to $H(x|z)$ when z is a continuous random variable and then the martingale convergence theorem will yield that:

$$H(\{x_j\}_1^\infty) = H(x_1|x_2, \, x_3, \, \ldots \, x_n, \, \ldots \,)$$

where the infinite sequence is treated as a continuous random variable.

For some simple examples consider:

a) i.i.d. processes. If the $\{x_j\}_1^\infty$ are independent and identically distributed then the entropy of the process is the entropy of the one-dimensional distribution.

b) Markov chains. If the $\{x_j\}_1^\infty$ form a Markov chain with transition probabilities p_{ij} and stationary measure π_i then the entropy is given by

$$-\sum_{i,\,j} \pi_i\, p_{ij}\, \log p_{ij}.$$

It was Kolmogorov's wonderful idea to use this entropy to give a numerical invariant to a measure preserving process as follows:

$$H(X,\ \mathcal{B},\ \mu,\ T) = \sup_{\mathcal{P}}\ H(T,\ \mathcal{P})$$

where we are using the notation $H(T,\ \mathcal{P})$ for the entropy of the stationary process defined by the partition $\mathcal{P}$, or in other words

$$H(T,\ \mathcal{P}) = \lim_{n\to\infty} \frac{1}{n}\, H(\mathcal{P} \vee T^{-1}\mathcal{P} \vee \ldots T^{-n+1}\mathcal{P}).$$

This is clearly invariant under measure preserving isomorphisms and was used by Kolmogorov to distinguish between Bernoulli shifts of different entropy such as the two shift and three shift, which are i.i.d. with distributions $(\frac{1}{2},\ \frac{1}{2})$, $(\frac{1}{3},\ \frac{1}{3},\ \frac{1}{3})$ respectively.

An important class of transformations defined by this entropy are the zero entropy ones. For a two-sided process the entropy can also be given as

(6) $H(x_0|x_{-1},\ x_{-2},\ x_{-3},\ \ldots x_{-n},\ \ldots\,) = H(T,\ \mathcal{P})$

and in this form a zero entropy process is seen to be characterized by the fact that x_0 is measurable with respect to the σ-field generated by the past $\{x_n\}_{-\infty}^{-1}$. This kind of prediction gives the name **deterministic** to the zero entropy processes. Thus a system $(X,\ \mathcal{B},\ \mu,\ T)$ has zero entropy if and only if all of the processes defined by it are deterministic. Many classically defined systems such as the rotation of the circle by an angle α, or more generally any translation on a compact group, belong to this class.

At the other extreme lie the transformations with **completely positive entropy** (c.p.e.) which are those for which no nontrivial partition gives rise to a deterministic process. If $\mathcal{P}$ is a generator, i.e.

$$\bigvee_{-\infty}^{\infty} T^{-m}\, \mathcal{P} = \mathcal{B}$$

or in other words X can be identified as the space of the process defined by $\mathcal{P}$, then the c.p.e. property can be characterized by the triviality of the remote past. This means that

$$\bigcap_{n=1}^{\infty} \bigvee_{k \geq n} T^{-k}\, \mathcal{P}$$

has only sets of measure zero or one. In probability theory, for the process $\{x_k\}_{-\infty}^{\infty}$, this is known as a zero-one law. A consequence of this alternate description of c.p.e. processes is the following.

It follows from the definitions that the entropy of a process, and of its time reversal

$$\hat{x}_n = x_{-n}$$

is the same. For transformations this implies that $(X, \mathcal{B}, \mu, T)$ is c.p.e. if and only if $(X, \mathcal{B}, \mu, T^{-1})$ is c.p.e. That implies the following:

Fact: The remote past is trivial if and only if the remote future is trivial.

This fact, while not involving entropy in an explicit way, seems to require the concept of entropy and its basic properties for its proof. In any event I do not know of a direct proof of this fact. More generally, in any system there is a σ-algebra $\mathcal{D} \subset \mathcal{B}$ that contains those sets A such that the partition $\{A, X \backslash A\}$ defines a deterministic process. This σ-algebra is clearly T-invariant and defines a factor of $(X, \mathcal{B}, \mu, T)$ that is the maximal deterministic factor. It was discovered by Pinsker, Rohlin and Sinai that if $\mathcal{P}$ is a generator for the system then this factor is the remote past of $\mathcal{P}$, or equivalently the remote future of $\mathcal{P}$. Thus the fact above is a special case of the following basic property of finite-valued stationary stochastic process $\{x_j\}_{-\infty}^{\infty}$, namely that modulo null sets one always has the equality of σ-fields:

$$\bigcap_{n=1}^{\infty} \mathcal{F}(\{x_j\}_{j=n}^{\infty}) \;=\; \bigcap_{n=1}^{\infty} \mathcal{F}(\{x_{-j}\}_{j=n}^{\infty}).$$

A more combinatorial interpretation of the Shannon entropy is given by the Shannon-McMillan theorem which I shall formulate for ergodic processes in several ways.

SHANNON-MCMILLAN THEOREM (SM). *For an ergodic stationary finite valued* $(\{1, 2, \ldots a\} = A)$ *process* $\{x_j\}$ *with entropy equal to* h, *for any* $\epsilon > 0$ *there is an* n_0 *and for all* $n \geq n_0$, *there is a set of outcomes* $T_n \subset A^n$ *satisfying:*

(i) $P\{(x_1, \ldots x_n) \in T_n\} \geq 1 - \epsilon$

(ii) for any $t \in T_n, \exp(-n(h + \epsilon)) \leq P((x_1, \ldots x_n) = t) \leq \exp(-n(h - \epsilon))$.

In this formulation we say that most of the outcomes $t \in A^n$ have approximately the same probability and this gives rise to the name **asymptotic equipartition property** (AEP) in information theory. Another form of the theorem which is better suited for certain generalizations speaks about how many outcomes in A^n are needed to cover a $(1 - \epsilon)$ portion of the probability space and then assertion is roughly as follows for $\epsilon > 0$, and n sufficiently large:

One can cover $1 - \epsilon$ **of the probability space by a set of outcomes** $T_n \subset A^n$ **of cardinality less than** $\exp n(h + \epsilon)$, **but one cannot cover even an** ϵ **fraction of the space by a set of outcomes of cardinality less than** $\exp n(h - \epsilon)$.

It is not hard to go from one formulation to the other. The second one makes it possible to give a "counting" definition of the entropy as follows: Fix some $0 < b < 1$. For each n, let t_n be the minimal cardinality of a set of outcomes in A^n whose total probability is at least b. Then

$$h = \overline{\lim_{n \to \infty}} \frac{1}{n} \log t_n.$$

For ergodic processes it turns out that the limit exists and is independent of b. Proofs of these facts can be found in many standard texts.

There is also a pointwise version of this theorem which we shall discuss later on (see chapter 9.)

The radically different behavior of the deterministic systems and systems with completely positive entropy can be made very precise by means of the concept of **disjointness** introduced by H. Furstenberg [Fu-1967].

DEFINITION. Measure-preserving systems $(X_i,\ \mathcal{B}_i,\ \mu_i,\ T_i)$ $i = 1,\ 2$ are **disjoint** if whenever they are both factors of a third system $(X,\ \mathcal{B},\ \mu,\ T)$ via factor maps

$$\pi_i :\ X \to X_i\,; \qquad \pi_i T = T_i \pi_i\,; \qquad \pi_i \circ \mu = \mu_i$$

the σ-algebras $\pi_i^{-1}(\mathcal{B}_i) \subset \mathcal{B}$ are independent.

We already encountered an example of this idea in chapter 3 where we saw that if no multiple of α is in the point spectrum of $(X,\ \mathcal{B},\ \mu,\ T)$ then $\mathcal{X} = (X,\ \mathcal{B},\ \mu,\ T)$ is disjoint from the rotation by α. In particular, the weakly mixing transformations, having no point spectrum at all, are disjoint from all circle rotations and more generally from all compact group rotations.

THEOREM 7.1. *Any zero entropy system is disjoint from any c.p.e.*

Before proving the theorem we will provide yet another characterization of c.p.e. systems. In the course of obtaining this characterization we shall need formula (7) below which will be a big ingredient in the proof of the theorem.

PROPOSITION 7.2. *A system $\mathcal{X} = (X,\ \mathcal{B},\ \mu,\ T)$ has completely positive entropy if and only if for any partition $\mathcal{P}$ of X,*

$$\lim_{n \to \infty}\ H(T^n,\ \mathcal{P}) = H(\mathcal{P}).$$

PROOF. It follows easily from the definition that if $H(T,\ \mathcal{Q}) = 0$ then also

$$H(T^n,\ \mathcal{Q}) = 0$$

for all n, so that the condition is clearly necessary. The sufficiency follows easily from the triviality of the remote past since that gives, via the Martingale convergence theorem, that

$$H(\mathcal{P}\ |\ \bigvee_{-\infty}^{-n}\ T^{-j}\,\mathcal{P}) \to H(\mathcal{P})$$

which certainly implies that

$$H(\mathcal{P}|\ \bigvee_{k=-\infty}^{-1}\ T^{-kn}\,\mathcal{P}) \to H(\mathcal{P}).$$

To see that c.p.e. implies that the remote past is trivial one has to verify one direction of the Pinsker-Rohlin-Sinai theorem that I alluded to earlier, namely one has to see that any set A that is measurable with respect to the remote past defines a deterministic process.

For any two stochastic processes $\{x_j\}$, $\{y_j\}$ writing out the pairs $(x_0,\ y_0)$, $(x_{-1},\ y_{-1})$, ... $(x_{-n},\ y_{-n})$ in the order $(x_0,\ x_{-1},\ ...\ x_{-n},\ y_0,\ y_{-1},\ ...\ y_{-n})$ using (5), and passing to the limit as n tends to infinity one gets from the martingale convergence theorem:

$$(7) \qquad H(\{x_j,\ y_j\}) = H(x_0\ |\ x_{-1},\ ...\ x_{-n},\ ...\ ,\ \{y_j\}_{-\infty}^{\infty})$$
$$+ H(y_0\ |\ y_{-1},\ y_{-2},\ ...\ y_{-n},\ ...\ ).$$

On the other hand, it is easy to see that the entropy of the pair process $(x_j,\ y_j)$ is also given for any n by

$$(8) \qquad \frac{1}{n}\ H(\{x_j,\ y_j\}_1^n\ |\ \{x_j,\ y_j\}_{-\infty}^0)$$

and calculating (7) using the conditional form of (5) and the same order as before gives the following expression for the entropy of the pair process:

$$(9) \qquad H(x_0|\{x_j\}_{-\infty}^{-1}\ \vee\ \{y_j\}_{-\infty}^{+\infty}) + \lim_{n\to\infty}\ H(y_0|\{y_j\}_{-\infty}^{-1}\ \vee\ \{x_j\}_{-\infty}^{-n})$$

One uses now the equality of (7) and (9) to deduce:

$$(10) \qquad H(y_0|y_{-1},\ y_{-2},...y_{-n},...) = \lim_{n\to\infty}\ H(y_0|\{y_j\}_{-\infty}^{-1}\ \vee\ \{x_j\}_{-\infty}^{-n})$$

for any process $\{x_j\}$. Taking now for $\{x_j\}$ the generator of a system and for $\{y_j\}$ a process defined by a set in the remote past of the generator, we see that the right-hand side of (10) vanishes which shows that any process defined by sets in the remote past is deterministic. This completes the proof of the proposition. $\qquad \square$

PROOF OF THEOREM 7.1. It suffices to show that if $\{x_n\}$ is c.p.e. and if $\{y_n\}$ is zero entropy, then no matter how they are joined the processes are independent. For this it will suffice to show that x_0 is independent of any finite collection $\{y_j\}_{-M}^M$. If, for some M, this independence would fail we would have by (4) that

$$(11) \qquad H(x_0|y_{-M},...y_M) < H(x_0).$$

We call the $(2M+1)$-tuples $\ \bar{y}_n\ =\ (y_{n-M}, y_{n-M+1}, \ldots, y_n, \ldots, y_{n+M})$ and consider the pair process $(x_n, \bar{y}_n)$ which is also stationary.

Now for N large enough, the proposition tells us that the stationary process defined by $\{x_{jN}\}_{j=-\alpha}^{\infty}$ would have process entropy very close to $H(x_0)$. But now formula (7) applied to the $\{\overline{y}_{jN}\}$ process and the $\{x_{jN}\}_{j=-\alpha}^{\infty}$ process will lead to a contradiction since the right hand side will be strictly less than $H(x_0)$ by some definite amount given by (11). $\square$

We will use this disjointness in the next chapter in our study of normal numbers and selection rules. Here we will give a simple application of the disjointness concept to unique ergodicity.

PROPOSITION 7.3. *Suppose that (Y, T) is uniquely ergodic with the unique invariant measure μ. Let $\pi_i : Y \to X_i$ be factor maps, so that $\pi_i T = T_i \pi_i$ for homeomorphisms T_i of X_i and suppose that for any non-empty open sets $U_i \subset X_i$ one has*

$$\mu(\pi_1^{-1}(U_1) \cap \pi_2^{-1}(U_2)) > 0.$$

Then the systems $(X_i, T_i, \pi_i \circ \mu = \mu_i)$ are disjoint.

PROOF. Consider the mapping $\pi : Y \to X_1 \times X_2$ defined by

$$\pi(y) = (\pi_1(y), \pi_2(y)).$$

Since π is continuous $\pi(Y)$ is a closed subset of $X_1 \times X_2$. The hypothesis implies that $\pi(Y)$ is dense and therefore equals all of $X_1 \times X_2$. If the systems $\mathcal{X}_i$ would not be disjoint there would be at least two distinct $T_1 \times T_2$ invariant measures on $X_1 \times X_2$, namely the product measure $\mu_1 \times \mu_2$ and some other measure. But then via π these could be lifted to distinct measures on Y contradicting the unique ergodicity of (Y, T). $\square$

COROLLARY 7.4. *A diagram of the form*

of ergodic measure theoretic systems with $\mathcal{X}_i$'s independent in $\mathcal{Y}$ and the $\mathcal{X}_i$'s not disjoint cannot have a uniquely ergodic model.

An example of this would be $\mathcal{X}_1 = \mathcal{X}_2$ any weakly-mixing system with $\mathcal{Y} = \mathcal{X}_1 \times \mathcal{X}_2$. This corollary should be contrasted with the following theorem (cf [W-1985]):

THEOREM 7.5. *Any diagram of the form*

of ergodic measure theoretic systems has a uniquely ergodic model.

There are topological analogues of entropy and disjointness which we shall now describe. If X is a compact space and $\mathcal{U}$ is an open cover of X we denote by $N(\mathcal{U})$ the minimal cardinality of a subset of $\mathcal{U}$ whose union is all of X. The compactness of X ensures that $N(\mathcal{U})$ is always finite. If $\mathcal{U}_1, \ldots \mathcal{U}_k$ are open covers we denote by

$$\mathcal{U}_1 \vee \mathcal{U}_2 \vee \ldots \vee \mathcal{U}_k = \{U_1 \cap U_2 \cap \ldots U_k : U_i \in \mathcal{U}_i \ 1 \leq i \leq k\}$$

the **join** of these covers. In analogy with Shannon's entropy, R. Adler, A. Konheim and H. McAndrew defined

$$H(T, \mathcal{U}) = \lim_{n \to \infty} \frac{1}{n} \log N \left(\bigvee_o^{n-1} T^{-j} \mathcal{U} \right)$$

where $\mathcal{U}$ is a cover of X and T a continuous mapping. The existence of the limit follows from the easily checked submultiplicative property of N:

$$N(\mathcal{U}_1 \vee \mathcal{U}_2) \leq N(\mathcal{U}_1) \cdot N(\mathcal{U}_2).$$

To get an invariant under topological conjugacy one defines the **topological entropy** of (X, T) to be

$$h_{\text{top}}(X, T) = \sup_{\mathcal{U}} H(T, \mathcal{U}).$$

To get some feeling for the definition consider a symbolic system, i.e. a closed space $X \subset A^{\mathbb{Z}}$ with $A = \{1, 2, \ldots a\}$ finite, $T = $ shift and $TX = X$. For $\mathcal{U}$ we can take the cover defined by the a different values the 0th coordinate of x can be. One sees then that $H(T, \mathcal{U})$ has a combinatorial interpretation as the logarithmic size of the allowable $n-$words. More explicitly let

$$X_n = \{\omega \in A^n : \text{ for some } x \in X, \quad x_j = \omega_j, \ 1 \leq j \leq n\}$$

and then we have

$$H(T, \mathcal{U}) = \lim_{n \to \infty} \frac{1}{n} \log |X_n|.$$

It is not hard to see that this is in fact the topological entropy of the system (X, T). For this example one sees easily that for any invariant measure μ

$$(12) \qquad\qquad h_\mu(X, T) \leq h_{\text{top}}(X, T).$$

Indeed this is always the case. A proof of this fact can be given along the following lines. First of all one shows that the entropy is an affine function on the space of invariant measures. This means that it suffices to consider ergodic measures in (12). Next one takes a partition on $\mathcal{P} = \{P_1, \ldots P_k\}$ into k−sets such that $H(T, \mathcal{P})$ is close to the entropy of (X, T, μ) and proceeds to construct an open cover $\mathcal{U}$ as follows.

Fix some small $\epsilon > 0$, and find disjoint closed sets $F_j \subset P_j$ that exhaust all but ϵ of the μ−measure of the space. Then take **disjoint** open sets U_j that contain the F_j and let U_0 be an open set that contains $X \setminus \bigcup_1^k U_j$ and is disjoint from $F = \bigcup_1^k F_j$. I claim that for the open cover $\mathcal{U}$ consisting of these U_j and U_0 the topological entropy cannot be much smaller than the measure theoretic entropy $h = H(T, \mathcal{P}, \mu)$. Indeed for large n, and for most of the points of the space X we have that the number of $i \in [0, n)$ for which $T^i x$ is not in $\bigcup_1^k F_j$ is at most $2\epsilon n$. This follows at once from the ergodic theorem. We now show by contradiction that $H(T, \mathcal{U})$ cannot be much less than h. If it were, say $h' < h$, then for all large n, there would be a subcover of $\bigvee_0^{n-1} T^{-j}\mathcal{U}$ of cardinality essentially $\exp nh'$. If $T^j x \in F = \bigcup_1^k F_i$ then $T^j x \notin U_0$ and then the $\mathcal{U}$−name determines the $\mathcal{P}$−name. Thus dividing up the $2\epsilon n$ fraction into all possible $\mathcal{P}$−names we still get definitely less than $\exp nh$ for the total number of different $\mathcal{P}$−names for all of these points and this is in conflict with the Shannon-McMillan theorem.

Although I have not gone into the details I believe that I have given the main ideas needed to prove:

PROPOSITION 7.6. *For a compact dynamic system (X, T) one has for all $T-$invariant measures μ:*

$$h_\mu(X, T) \leq h_{\text{top}}(X, T).$$

Here we are making explicit the dependence on μ in the measure theoretic entropy of the transformation. In full generality this was established by W. Goodwynn. A little bit later I. Dinaburg and T.N.T. Goodman showed that one actually has

$$(13) \qquad \sup_\mu \ h_\mu(X, T) = h_{\text{top}}(X, T).$$

A more perspicacious proof involving a metric definition of the topological entropy was given shortly afterward by R. Bowen. Quite recently, F. Blanchard, B. Host and E. Glasner have given a new proof of the variational principle (12) with a sharper statement. I would like to give a brief account of this work since it adds an interesting combinatorial element to the open covers of topological entropy. Their theorem is as follows:

THEOREM 7.7. *If T is a homeomorphism of a compact metric space X, and $\mathcal{U}$ is an open cover of X, then there is an invariant measure μ such that for any borel measurable partition $\mathcal{P}$ that is subordinated to $\mathcal{U}$ we have*

$$(14) \qquad H_\mu(T, \mathcal{P}) \geq H_{\text{top}}(T, \mathcal{U}).$$

Here, $\mathcal{P}$ **subordinated to** $\mathcal{U}$ means that each atom P of $\mathcal{P}$ is contained in some $U \in \mathcal{U}$.

The strategy of the proof is to prove the theorem in the special case that X is $0-$dimensional. Then one obtains the general case as follows: Any (X, T) has a $0-$dimensional extension $\pi : \widehat{X} \to X$ with a homeomorphism $\widehat{T}$ of $\widehat{X}$ such that

$$\pi\widehat{T} = T\pi.$$

One establishes this just like one shows that any compact metric space is a quotient of the Cantor set. Next one defines $\widehat{\mathcal{U}}$ as $\pi^{-1}(\mathcal{U})$ and applies the theorem there to find a $\hat{\mu}$ satisfying

$$H_{\hat{\mu}}(\widehat{T}, \widehat{\mathcal{P}}) \geq H_{\text{top}}(\widehat{T}, \mathcal{U}).$$

Now for any $\mathcal{P}$ subordinate to $\mathcal{U}$, $\pi^{-1}(\mathcal{P})$ is a partition of $\widehat{X}$ subordinated to $\widehat{\mathcal{U}}$. Furthermore even though in general the entropy may increase in going to an extension, for partitions $\widehat{\mathcal{P}}$ of the form $\pi^{-1}(\mathcal{P})$ we have

$$H_{\hat{\mu}}(\widehat{T},\ \widehat{\mathcal{P}}) = H_\mu(T,\ \mathcal{P})$$

and thus the conclusion of the theorem in the general case follows from the special case. Note that $H_{\mathrm{top}}(\widehat{T},\ \widehat{\mathcal{U}}) = H_{\mathrm{top}}(T,\ \mathcal{U})$.

The next reduction says that it suffices to consider partitions into closed and open sets since in a 0-dimensional space partitions of this form subordinated to $\mathcal{U}$ are dense in the space of all partitions subordinated to $\mathcal{U}$, with respect to the metric on partitions defined for fixed μ by:

$$\rho(\mathcal{P}_1,\ \mathcal{P}_2) = H_\mu(\mathcal{P}_1|\mathcal{P}_2) + H_\mu(\mathcal{P}_2|\mathcal{P}_1).$$

In this metric $H_\mu(T,\ \mathcal{P})$ is a Lipschitz function, as V. Rohlin observed, and therefore we may restrict to such nice partitions.

Now, as usual, invariant measures will be found as cluster points in the w^*-topology of a sequence of measures λ_n defined as follows:

$$\int f(x)d\lambda_n(x) = \frac{1}{N_n} \sum_0^{N_n-1} f(T^j x_n), \qquad f \in C(X).$$

The N_n will be tending to infinity and the x_n will be chosen shortly. The form of λ_n guarantees that any such cluster point is an invariant measure. Let us fix a single partition $\mathcal{P}$ into closed and open sets that is subordinate to $\mathcal{U}$ and show how to construct a measure μ such that (14) will hold for this $\mathcal{P}$. The property of x_n that we will need is the following. Fix some sequence ϵ_k that tends to zero and let $h = H_{\mathrm{top}}(T,\ \mathcal{P})$. For all $k \leq n$ we want x_n to have the property that the empirical distribution on k-blocks given by the $\mathcal{P} - N_n$ name of x_n, say $q_{n,k}$ satisfies

$$(15) \qquad\qquad \frac{1}{k} H(q_{n,k}) \geq h - \epsilon_k.$$

Explicitly, if $\mathcal{P} = \{P_1,\ P_2, \ldots P_a\}$, $\{1,\ 2, \ldots a\} = A$ then the $\mathcal{P} - N_n$ name of x_n is the element $\alpha \in A^{N_n}$ such that for all $0 \leq i < N_n$

$$T^i x_n \in P_{\alpha(i)}.$$

The distribution $q_{n,k}$ is defined by setting:

$$q_{n,k}(\beta) = \frac{|\{0 \leq i < N_n - k : \alpha(i)\, \alpha(i+1) \ldots \alpha(i+k-1) = \beta(0) \ldots \beta(k-1)\}|}{N_n - k}$$

This is a probability distribution on A^k.

The sets defined by (15) are closed and it is easy to see that if the x_n's satisfy these inequalities for all $k \leq n$ then any cluster point of the λ_n's will indeed satisfy (14) as required. Now comes the main idea - namely that points x_n with these properties exist for any N_n sufficiently large. This comes about because the total number of N_n-names in A^{N_n} that satisfy the reverse inequality

(16) $$\frac{1}{k}\, H(q_{n,k}) < h - \epsilon_k$$

can be estimated from above by

$$\exp(N_n \cdot (h - \delta_k))$$

for some $\delta_k > 0$ that depends on ϵ_k. This is a combinatorial fact that is well known in information theory. I will return to this point after showing how to use it. For any N, the meaning of h is that no subcover of $\bigvee_0^{N-1} T^{-i}\, \mathcal{U}$ can have fewer than $\exp N_n \cdot h$ elements. Now if there would be no point satisfying (14), the total number of $\mathcal{P} - N_n$ names of points in the space would be too small and therefore, for N_n sufficiently large points x_n with the desired properties must exist.

The fact that $\mathcal{P}$ is a partition into closed and open sets means that μ−measures of the atoms of $\mathcal{P} \vee T^{-1}\mathcal{P} \vee \ldots T^{-k+1}\mathcal{P}$ can be calculated from the defining formulae for λ_n since their indicator functions are continuous. Modulo the combinatorial estimate on names this completes the proof for a single partition. To deal with all partitions into closed and open sets one observes that the collection is countable and at stage n, finds an x_n with the above properties for the first n−partitions in an exhaustive list. Since there is an exponential factor in the combinatorial estimate this is possible. Finally a word about the combinatorial estimate. We will content ourselves with explaining the case $k = 1$.

Fix a distribution on the symbol set $\{1, 2, \ldots a\} = A$ say $(q_1, q_2, \ldots q_a)$. Put the product measure q^N on A^N, i.e. independent random variables. Now any

$\alpha \in A^N$ whose empirical 1–block distribution is $(\frac{b_1}{N},\ \frac{b_2}{N}, \ldots \frac{b_a}{N})$ carries weight

$$\prod_{j=1}^{a} q_j^{b_j} = \left(\prod_{j=1}^{a} q_j^{\frac{b_j}{N}}\right)^N.$$

If this empirical distribution is within δ of the q_j's, i.e.

$$\sum_{1}^{a} |q_j - \frac{b_j}{N}| < \delta.$$

Then this weight is greater than or equal to

$$\left(\prod_{j=1}^{a} q_j^{q_j}\right)^N \cdot \left(\prod_{j=1}^{a} q_j\right)^{-\delta N} = \exp\{-[H(q) + \delta \cdot c]N\}$$

and therefore the number of such α's is at most the reciprocal of this expression. Since there is an exponential involved and the total number of distributions q_j's in any δ-net in the ℓ_1-metric is fixed there is no difficulty in putting together all of these estimates to yield the fact that we needed above. To be sure the δ needed for this δ-net is not the same as the δ in our discussion above.

One application of this theorem is to give conditions that guarantee that a system (X, T) has a measure realizing the maximal entropy. It follows from the theorem that a sufficient condition is the existence of a finite cover $\mathcal{U}$ realizing the topological entropy. A sufficient condition for that to happen is for $\mathcal{U}$ to separate points in the sense that the maximum diameter of elements of $\bigvee_0^{N-1} T^{-i} \mathcal{U}$ tends to zero with n.

Topological disjointness of two systems $(X_i,\ T_i)$, $i = 1,\ 2$ is simply the requirement that whenever there is a common topological extension $(Y,\ T)$ with equivariant maps $\pi_i : Y \to X_i$, one can factor these maps through the direct product $(X_1 \times X_2,\ T_1 \times T_2)$. This means that there is an equivariant onto mapping $\rho : Y \to X_1 \times X_2$ so that the π_i are given by the composition of ρ with the coordinate projections. It is not hard to see that for disjointness to take place at least one of the systems **must** be minimal and in discussing disjointness one usually supposes that both systems are minimal.

The simple obstruction to disjointness is the existence of a common factor for the systems $(X_i,\ T_i)$. Indeed, if $(Z,\ S)$ is a factor of both of them with maps

$\sigma_i : X_i \to Z$, then the fiber product

$$X_1 \underset{Z}{\times} X_2 = \{(x_1,\ x_2) : \sigma_1(x_1) = \sigma_2(x_2)\}$$

will be a common extension of the X_i's that does not factor through the full direct product. A similar condition holds for the measure theoretic disjointness.

For simple examples of topological disjointness we have again the rotations on compact groups – now thought of as topological systems – which are disjoint from the topologically weak mixing transformations. Here one sees that not having common factors implies disjointness. In general this does not hold. Using horocycle flows on compact surfaces we [GW] were able to give examples of transformations without common factors which are not disjoint. Subsequently several other examples of this type of phenomenon were given.

Recently, E. Lindenstrauss was able to show a rather different route to this phenomenon which is closely connected to entropy. He showed that if one looks at the coordinate shift in the Hilbert cube $[0,\ 1]^{\mathbb{Z}}$ as a topological system then any nontrivial factor of it must have infinite entropy. He went on to construct an example of a minimal system $(Z,\ S)$ with the same property, namely that any nontrivial factor of it has infinite entropy.

Now let $(\widehat{Z},\ \widehat{S})$ be a minimal zero-dimensional extension of $(Z,\ S)$. As $\widehat{Z}$ is zero dimensional it has nontrivial factors in symbolic systems with a finite number of symbols, and in fact these generate the entire system. These factors have finite entropy and are not disjoint from $(Z,\ S)$, but have no factor in common with it.

The nondisjointness can be made quite dramatic as follows. An extension

$$\pi : Y \to X$$

is called **almost one to one** (or almost 1-1) if there is a dense G_δ of points x for which $\pi^{-1}(x)$ is a single point. For minimal systems this follows from the existence of a single x_0 with $|\pi^{-1}(x_0)| = 1$ since the set $\{x \in X : \pi^{-1}(x)| = 1\}$ is always a G_δ. This comes about because the set

$$\{x : \mathrm{diam}(\pi^{-1}(x)) \geq b > 0\}$$

is closed (recall that Y is compact) and

$$\{x : |\pi^{-1}(x)| = 1\} = \bigcap_{K=1}^{\infty} \left\{x \in X : \operatorname{diam}(\pi^{-1}(x)) < \frac{1}{K}\right\}.$$

In a certain topological sense almost 1-1 extensions are not distinguishable from the system itself. Now both factors of $(\widehat{Z}, \widehat{S})$ above can be made almost 1-1. That is, we can find an almost 1-1 extension $(\widehat{Z}, \widehat{S})$ of (Z, S) that is zero dimensional, and then we can find a factor (X, T) of $(\widehat{Z}, \widehat{S})$ with finite entropy such that $\widehat{Z}$ is an almost 1-1 extension of (X, T). In spite of the fact that in a generic sense X and Z are indistinguishable, they have no common factor.

References

1. R.L. Adler, A.G. Konheim and M.H. McAndrew, *Topological entropy*, Trans. Amer. Math. Soc. **114**(1965), 309–319.
2. R. Bowen, *Entropy for group endomorphisms and homogeneous spaces*, Trans Amer. Math. Soc **153**(1971), 401–413.
3. F. Blanchard, E. Glasner and B. Host, *A variation on the variational principle and applications to entropy pairs*, Ergodic Theory Dynam. Systems **17**(1997), 29–43.
4. E.I. Dinaburg, *The relation between topological entropy and metric entropy*, Dokl. Akad. Nauk SSSR **190**(1970) = Soviet Math. Dokl. **11** Nr. 1(1970), 13–16.
5. H. Furstenberg, *Disjointness in ergodic theory*, Math. Syst. Theory **1**(1967), 1–49.
6. T.N.T. Goodman, *Relating topological entropy with measure theoretic entropy*, Bull. London Math. Soc. **3**(1970), 176–180.
7. L. Goodwynn, *Topological entropy bounds measure theoretic entropy*, Proc. Amer. Math. Soc. **23**(1969), 679–688.
8. S. Glasner and B. Weiss, *Minimal transformations with no common factor need not be disjoint*, Israel J. of Math., **45**(1983), 1–8.
9. E. Lindenstrauss, *Lowering topological entropy*, J. d'Analyse Math. **67**(1995), 231–267.
10. B. Weiss, *Strictly Ergodic Models for Dynamical Systems* BAMS (NS) **13**(1985), 143–146.

CHAPTER 8

What is Randomness?

The title question of this chapter has been the subject of many investigations in disciplines ranging from philosophy and psychology to computer science and physics. I shall not attempt to give even a cursory survey of the many insights that have been attained and will confine myself to some mathematical answers that are related to our theme of single orbit dynamics. For our first attempt at understanding randomness, consider the opposite of randomness - that is **determinism**. In the previous chapter we saw that for stochastic processes determinism is associated with zero entropy and thus positive entropy can be thought of as entailing some randomness. Indeed this is so in the following sense: If a measure preserving system $(X, \mathcal{B}, \mu, T)$ has positive entropy then there is a nontrivial partition $\mathcal{P}$ of X such that the process defined by $\mathcal{P}$ consists of independent random variables, i.e. the partitions $\{T^{-i}\mathcal{P}\}_{i \in \mathbb{Z}}$ are independent. One can choose $\mathcal{P}$'s to have any abstract distribution q satisfying

$$H(q) \le h(X, \mathcal{B}, \mu, T).$$

This result of Y. Sinai established the weak isomorphism of Benoulli shifts of the same entropy. The isomorphism itself was established several years later by D. Ornstein in his path breaking [O-1970] which begins with a fresh proof of this fundamental result.

I would like to explain a combinatorial-topological version of this idea that positive entropy means randomness. The result deals with symbolic systems $X \subset \{0,1\}^{\mathbb{Z}}$, that are closed and shift invariant. The randomness of X will be "measured" by the size of the **interpolating sequences** (or **interpolating sets**) of X. The idea is that the full 2-shift is as random as possible since anything that can happen may actually occur. A subset $I \subset \mathbb{Z}$ is called an **interpolating set** for X

if

$$X|_I = \{0, 1\}^I$$

or, more explicitly, for any $\alpha \in \{0, 1\}^I$ there is some $x \in X$ such that

$$x(i) = \alpha(i) \text{ for all } i \in I.$$

If X has interpolating sets of positive density then clearly X will have positive topological entropy. This follows immediately from the identification of the topological entropy of X as the growth rate of the number of n-blocks that one sees in X. The surprising fact is that this is the only reason for X to have positive entropy. Indeed the following theorem is true:

THEOREM 8.1. *If $X \subset \{0, 1\}^{\mathbb{Z}}$ is closed and shift invariant then it has non-zero topological entropy under the shift if and only if it has interpolating sets of positive density.*

At the heart of the proof lies a combinatorial fact that was proved independently by N. Sauer and M. Perles (see [Shelah]). This fact is a precise estimate for the finite version of the problem. We deal with subsets B of the binary cube $\{0, 1\}^n$ and ask for the maximum cardinality of a subset $B \subset \{0, 1\}^n$ that does **not** have an interpolating set of cardinality k. The simplest way that we can force B not to have an interpolating set of size k is to make sure that B contains no element ξ of $\{0, 1\}^n$ with k or more ones, because then no matter what $I \subset n$ we would choose, the element $\alpha(i) = 1$ for all $i \in I$ could not be interpolated. Now there are precisely

$$\sum_{j=0}^{k-1} \binom{n}{j}$$

such elements ξ. The fact alluded to above is that this is the best that one can do, i.e.

PROPOSITION 8.2. *(N. Sauer - M. Perles): If $B \subset \{0, 1\}^n$ satisfies*

$$|B| > \sum_{j=0}^{k-1} \binom{n}{j}$$

then B has an interpolating set with k-elements.

For completeness we include a proof of this nice combinatorial fact. Indeed one can prove a stronger statement. For some collection $\mathcal{S}$ of subsets $S \subset \{1, 2, \ldots n\}$ suppose that to each $S \in \mathcal{S}$ is assigned some $\sigma_S \in \{0,1\}^S$. Let

$$B(\mathcal{S}, \sigma) = \{\alpha \in \{0,1\}^n : \alpha|_S \neq \sigma_S \text{ for all } S \in \mathcal{S}\}.$$

Let $B_0(\mathcal{S})$ denote the collection obtained when all the σ_S equal the constant -0 element. Then I claim that

$$|B_0(\mathcal{S})| \geq |B(\mathcal{S}, \sigma)|.$$

To see this it suffices to study what happens when σ is modified to $\overline{\sigma}$ by uniformizing in one place, i.e.

$$\overline{\sigma}_S(i) = \begin{cases} \sigma_S(i) & \text{for } i \neq 1 \\ 0 & \text{if } i = 1. \end{cases}$$

Note that changes are actually made only for those S's that contain 1. Since B_0 is obtained in n-steps of this type from $B(\mathcal{S}, \sigma)$ this clearly suffices. Now suppose $\alpha \in B(\mathcal{S}, \sigma)$ but $\alpha \notin B(\mathcal{S}, \overline{\sigma})$. Then for some S containing 1 we have $\alpha|_S = \overline{\sigma}_S$. Since $\alpha|_S \neq \sigma_S$ if we would change α to $\overline{\alpha}$ by putting $\overline{\alpha}(1) = 1$ and $\overline{\alpha}(i) = \alpha(i)$ for all $i \neq 1$, we would have $\overline{\alpha} \neq B(\mathcal{S}, \sigma)$ but $\overline{\alpha} \in B(\mathcal{S}, \overline{\sigma})$. This mapping being 1-1 we see that we have a bijection from $B(\mathcal{S}, \sigma) \backslash B(\mathcal{S}, \overline{\sigma})$ into $B(\mathcal{S}, \overline{\sigma}) \backslash B(\mathcal{S}, \sigma)$ which clearly implies

$$|B(\mathcal{S}, \overline{\sigma})| \geq |B(\mathcal{S}, \sigma)|$$

as required.

Applying this to the collection of all k-subsets of $\{1, 2, \ldots n\}$ gives the Sauer-Perles result.

Applying Stirling's formula to the binomial coefficients enables one to deduce from this the following corollary: There is a function b from the open interval $(0, 1)$ to itself so that for n sufficiently large, if $B \subset \{0,1\}^n$ satisfies:

$$|B| \geq 2^{hn}$$

then B has interpolating sets I of size:

$$|I| \geq b(h) \cdot n.$$

We now have to patch these finite interpolating sets into a single infinite one of positive density. For this some dynamical ideas will be useful.

Let us identify interpolating sets $I \subset \mathbb{Z}$ with their indicator functions and consider them too as points in $\{0,1\}^{\mathbb{Z}}$. Denote by $\mathcal{I}(X)$ the collection of such points. Thus $\mathcal{I}(X) \subset \{0,1\}^{\mathbb{Z}}$ and $y \in \mathcal{I}(X)$ if and only if for any $z \in \{0,1\}^{\mathbb{Z}}$ there is some $x \in X$ such that for all n:

$$y(n) \cdot x(n) = y(n) \cdot z(n).$$

The fact that X is closed implies that $\mathcal{I}(X)$ is closed as may be easily checked. Furthermore, since X is shift invariant so is $\mathcal{I}(X)$. Now the proposition gives us, for all sufficiently large n, sets $I \subset \{1,2,\ldots n\}$ with **density** at least $b(h)$ that are interpolating sets for $X_n \subset \{0,1\}^n$ where

$$X_n = X|_{\{1,2,\ldots n\}},$$

if the topological entropy of X equals $h > 0$. Using these finite sets one can define shift-invariant measure μ on $\mathcal{I}(X)$ that satisfies

$$\mu([1]) \geq b(h).$$

For this we note that not only is $\mathcal{I}(X)$ closed but it also contains the limit of any sequence of finite interpolating sets. Then the measure μ is constructed as a weak* cluster point of finite measures defined by the finite blocks corresponding to finite interpolating sets of positive density.

The pointwise ergodic theorem guarantees that there are points of positive density in $\mathcal{I}(X)$, since the averages of the visit times to $[1]$ will converge for $\mu - a.e.$ point to a function which is the projection of the indicator function $1_{[1]}$ on the space of invariant functions. Clearly for some points the limit will have to be at least $b(h)$. A similar result holds for larger alphabet sizes with an analogous proof. The ergodic theoretic part is unchanged while the combinatorial part must be generalized in the obvious fashion. Here is the result:

THEOREM 8.3. *If $X \subset \{1,2,\ldots a\}^{\mathbb{Z}} = A^{\mathbb{Z}}$ is closed and shift invariant with*

$$h_{\text{top}}(X) > \log(a-1)$$

then there is an interpolating set $I \subset \mathbb{Z}$ for X with positive density, the density depending only on the difference $h(X) - \log(a-1)$.

It is possible to use these results to give a characterization of completely positive entropy systems that has a topological flavor. For this let us associate to a system $(X, \mathcal{B}, \mu, T)$ and a process defined by a partition $\mathcal{P}$ a subshift $X_{\mathcal{P}} \subset A^{\mathbb{Z}}$, where $\mathcal{P} = \{P_1, \ldots P_a\}$, as follows: Map X to $A^{\mathbb{Z}}$ by assigning to each $x \in X$ its infinite $\mathcal{P}$-name $\theta(x)$:

$$\theta(x) = (\ldots \mathcal{P}(T^{-1}x), \ \mathcal{P}(x), \ \mathcal{P}(Tx), \ldots \mathcal{P}(T^n x), \ldots)$$

and set $X_{\mathcal{P}}$ to be the closed support of the measure $\theta \circ \mu$. That means that the allowed n-blocks of $X_{\mathcal{P}}$ correspond to atoms of $\bigvee_{0}^{n-1} T^{-i}\mathcal{P}$ that have positive measure. Observe that even very deterministic systems like rotations by irrational angles still admit processes P with $X_{\mathcal{P}}$ being the full shift $A^{\mathbb{Z}}$. Such processes can be easily constructed using Rohlin towers in a simple fashion - but we will not go into that now. However, we do have:

THEOREM 8.4. *A system* $(X, \mathcal{B}, \mu, T)$ *has completely positive entropy if and only if for all nontrivial* $\mathcal{P}$, $X_{\mathcal{P}}$ *has positive density interpolating sets.*

This theorem mixes the qualitative randomness expressed by large interpolating sets with a purely measure theoretic notion. The proof of the theorem goes as follows:

If $(X, \mathcal{B}, \mu, T)$ fails to have completely positive entropy then it has factors with zero entropy. As we just remarked this does not guarantee that the corresponding $X_{\mathcal{P}}$'s do not have positive entropy interpolating sets. However, any zero entropy system has some partitions $\mathcal{P}$ such that $X_{\mathcal{P}}$ has zero topological entropy. This follows easily from the proof of the Jewett-Krieger theorem or it can also be proved directly.

For the other direction, assume c.p.e. for the system. If $\mathcal{P}$ has only two sets then since the measure-theoretic entropy of $\theta \cdot \mu$ is positive the topological entropy of $X_{\mathcal{P}}$ is positive by the fact that topological entropy upper bounds the measure entropy. Now the theorem follows from theorem 8.1. If $\mathcal{P}$ has many sets then one cannot use theorem 8.3 directly because to begin with we only know that the topological entropy of $X_{\mathcal{P}}$ is positive. However, if we pass to T^N for some large N then as we saw in the preceding chapter the c.p.e. condition guarantees that the entropy of $(\mathcal{P}, \ T^N)$ will be close to $H(\mathcal{P})$. So once again we could complete the proof if the distribution of $\mathcal{P}$ was close enough to the uniform distribution, so that if $\mathcal{P}$ was a partition into b sets we would have

$$H(\mathcal{P}) > \log(b - 1).$$

However, in general this too does not hold so that to complete the proof the partition $\mathcal{P}$ has to be refined to a $\widehat{\mathcal{P}}$ so that the distribution of $\widehat{\mathcal{P}}$ is close to uniform. This is trivial if the distribution involves only rational numbers. For general distributions a careful approximation of the irrational numbers by rational numbers is needed. In both cases the final deduction is the observation that interpolating sets for $X_{\widehat{\mathcal{P}}}$ are certainly interpolating sets for $X_{\mathcal{P}}$. For details see [GW-1995].

We will now turn to randomness in the sense of von Mises and explore what can be said about these old classical notions using the techniques of single orbit dynamics. We shall explore to what extent the following thesis is valid:

THESIS. *Generic points for fair coin tossing are collectives.*

Recall that a point $x \in \{0, 1\}^{\mathbb{N}}$ is a generic point for the independent $(1/2, \ 1/2)$ measure if the asymptotic frequency of occurrence of any block of length k is exactly 2^{-k}. Thus these generic points are exactly the normal numbers (in base 2) of E. Borel. We also record the defining property of a **collective** for fair coin tossing:

COLLECTIVES.

(1) limiting frequencies of individual symbols exist.
(2) property (1) persists for any subsequence selected by an admissible selection rule.

If no restrictions are made on the nature of the selection rules then clearly (2) can never be satisfied, since we could always select out exactly those places where 0-occurred. Note too that using rather simple selection rules that should surely be admitted, such as select out all the places that immediately follow an occurrence of a zero, one can establish by induction the following lemma:

LEMMA 8.5. *Any sequence possessing the same limiting frequencies for all simple selection rules is generic for an independent process.*

By simple selection rules we mean all those that have the form: **select all places that follow the occurrence of a fixed block** w_0. The lemma explains why collectives have to be generic points. Our thesis is that generic points are collectives and what we shall now do it explore the set of selection rules that maps generic points into generic points. From the outset we shall restrict ourselves to nonanticipatory rules. That is we must decide whether or not to select x_n, the nth outcome, on the basis of what we have observed up to time $n - 1$. Formally, let $\{0, 1\}^*$ denote the collection of all finite words of zeros and ones, and let S be a subset of $\{0, 1\}^*$. Then S defines a selection rule as follows:

$$S(x) = x_{n_1} \, x_{n_2} \, x_{n_3} \ldots$$

where the n_i are the successive values of n such that

$$x_1 \, x_2 \, \ldots \, x_{n-1} \in S.$$

We shall adopt the convention that $S(x)$ is defined in this way only if the density of the resulting sequence is positive, otherwise we do not consider $S(x)$ to be defined. A corollary of our thesis is the following definition:

DEFINITION. A selection rule $S \subset \{0, 1\}^*$ is **admissible** if for all normal numbers x, $S(x)$ is a normal number.

We have now satisfied the requirements of von Mises, however, it remains to explore the nature of the set of admissible selection rules which is what we now undertake to do. It is worth repeating again that an **almost everywhere** result is not difficult to prove. In probabilistic language a selection rule defines a stopping time τ. If we assume that the stopping time has finite expectation, which is the probabilistic analogue of our convention that $S(x)$ is to be defined only if the density of the selected sequence is positive, then it is easy to see that the expected value of x_τ is the same as the expected value of x_1, and furthermore the successive stopping times $\tau_2, \ldots$ defined by restarting the process after X_{τ_1}, give independent random variables, so that the law of large numbers implies that almost surely the frequencies of 0 and 1 in (x_{τ_k}) is the same as $E(X_1)$.

This result, while certainly explaining why gambling systems do not work, is not satisfactory from our point of view. The exceptional null set clearly depends upon the stopping time τ.

The ensemble of stopping times is definitely not countable and this means that we cannot lump together all of the exceptional sets in one grand null set. As before, we are looking for results which will hold for **all** normal numbers. An example of such a result concerns the rules generated by finite state automata. Such objects, which are a general model for a finite computing device, consist in a set of states V, a function from $\{0, 1\} \times V$ to V denoted by σ, and two distinguished states, v_i and v_s, the initial state and the stopping state (which may be the same). A sequence $x_1 \, x_2 \ldots x_a$ is accepted by the automaton $(V, \, \sigma, \, v_i, \, v_s)$ if the sequence of states defined inductively:

$$v(0) = v_i$$

$$\vdots$$

$$v(m) = \sigma(x_m, \, v(m-1))$$

$$\vdots$$

$$v(a) = \sigma(x_a, \, v(a-1))$$

terminates with $v(a) = v_s$. Clearly the set of sequences accepted by an automaton $(V, \, \sigma, \, x_i, \, x_s)$ defines a selection rule.

THEOREM 8.6. *Any selection rule defined by a finite automaton is admissible.*

The collection of all finite automata is countable so that it would be easy to prove this theorem if we would only be concerned about the behavior of almost every normal number. For the proof of the theorem for all normal numbers we need an extension of the idea of unique ergodicity.

DEFINITION 8.7. A system (X, T) is said to be **intrinsically ergodic** if there us a unique invariant measure μ that maximizes the entropy of the system $(X, \, T, \, \mu)$.

The variational principle implies that a unique maximizing measure must achieve the topological entropy but that in itself is not sufficient to guarantee uniqueness. The basic example of an intrinsically ergodic system is the full shift

$A^{\mathbb{Z}}$, with $|A| = a$ finite. This property for the full shift follows immediately from the basic entropy inequality:

$$-\sum_{1}^{N} p_j \log p_j \leq \log N$$

with inequality only if all the p_j's equal $1/N$. This means that as soon as we deviate from uniformity in the distribution of k-blocks the entropy will drop below $\log |A|$ and it will never recover. An important generalization of this example are the shifts of finite type which are those symbolic systems that are defined by means of a finite list of forbidden blocks. Not all such systems are intrinsically ergodic. For example in $\{0,1\}^{\mathbb{Z}}$ if the block 10 is forbidden the system has just two invariant measures, the point masses concentrated on the fixed points. However, a simple transitivity condition, namely that it is possible to go from any finite configuration to any other, rules out this type of behavior and then the shifts of finite type are intrinsically ergodic. The proof of this was first given by W. Parry, another proof is found in [AW-1970]. This latter proof was vastly extended in a series of papers of R. Bowen to cover many smooth dynamical systems exhibiting hyperbolic behavior of different kinds.

We return now to theorem 8.6, and use the notation (V, σ, v_i, v_s) to denote a finite automaton. Let us define a shift of finite type in $(\{0,1\} \times V)^{\mathbb{Z}}$ by allowing only pairs $((\xi, v), (\bar{\xi}, \bar{v}))$ such that

$$\bar{v} = \sigma(\xi, v).$$

The topological entropy of the shift is $\log 2$, and the part of it that we are interested in, namely those sequences where v_s occurs with positive density, is clearly transitive, hence intrinsically ergodic. Now for any normal number $(\xi_n)_0^{\infty}$ we look at the point y in this shift of finite type generated by ξ_n, i.e. define inductively $(\xi_0, v_0) = (\xi_0, v_i)$, and then

$$(\xi_m, v_m) = (\xi_m, \sigma(\xi_{m-1}, v_{m-1})).$$

Any cluster point of the measures

$$\frac{1}{N} \sum_{j=1}^{N} \delta_{T^j y}, \qquad (T = \text{shift})$$

projects onto the independent measure on $\{0,1\}^{\mathbb{Z}}$ since (ξ_n) was assumed to be normal. Thus the entropy must be at least $\log 2$, since factors have no more entropy then the original transformation. By intrinsic ergodicity this measure is unique. If in a metric space a sequence has a unique cluster point it must actually converge to this limit. It now follows easily from the properties of the Markov chain that successive visits to v_s are independent and this will translate into the normality of the sequence selected by the finite automaton. Further details of this type of argument can be found in [KW-1974] where a more general result is established.

Perhaps the simplest class of selection rules are those that are independent of the input, i.e. S is defined by a sequence $\{s_1 < s_2 < \ldots\}$ and $S(x) = (x_{s_1}\ x_{s_2}\ x_{s_3}\ldots)$. For this class we can give exact necessary and sufficient conditions for the selection rule to be admissible. It is convenient to formulate the condition not in terms of the sequence $(s_1, s_2, \ldots)$ but rather in terms of its indicator function, that is to say the 0-1 sequence (Z_n) which is one exactly for those n's that equal one of the s_j's.

DEFINITION 8.8. A point $z \in \{0,1\}^{\mathbb{N}}$ is said to be **completely deterministic** if any cluster point of the measures

$$\frac{1}{N} \sum_1^N \delta_{T^j z}, \qquad (T = \text{shift})$$

defines a deterministic process, i.e. has zero entropy.

Implicit here is the assumption that the density of ones in z is positive. This is much weaker than requiring that the orbit closure of z has zero topological entropy. A combinatorial formulation is possible but a little cumbersome. Here is one possibility:

LEMMA 8.9. *a point $z \in \{0,1\}^{\mathbb{N}}$ is completely deterministic if and only if for any $\epsilon > 0$ there is some K such that after removing from $\{z_n\}_1^{\infty}$ a subset of density less than ϵ, what is left can be covered by a collection $\mathcal{C}$ of K-blocks such that*

$$|\mathcal{C}| < 2^{\epsilon K}.$$

That the condition of the lemma implies the complete determinism of z is straightforward. For the other direction one must use a more elaborate argument, (see for example van Lambalgen). The main result is the following:

THEOREM 8.10. *A point $z \in \{0,1\}^{\mathbb{N}}$ defines an admissible selection rule if and only if it is completely deterministic.*

One direction of the theorem follows pretty easily from the disjointness between independent processes and zero entropy processes. The argument goes as follows: Start with any sequence $m'_1 < m'_2 < \ldots$, and pass to a subsequence $\{m_j\}$ along which the measures

$$\frac{1}{m_j} \sum_{i=1}^{m_j} \delta_{T^i z}$$

converge to a limiting measure μ. Let $x \in \{0,1\}^{\mathbb{N}}$ be a normal sequence and look at the pair (z, x) in $(\{0,1\} \times \{0,1\})^{\mathbb{N}}$. Consider the measures

$$\frac{1}{m_j} \sum_{i=1}^{m_j} \delta_{T^i(z, x)}$$

where T denotes now the shift on the product space and pass to a subsequence that converges to a measure λ. We know the one-dimensional marginal distributions of λ, and since they correspond to disjoint processes (the μ-process has zero entropy by assumption) we know that λ is a product measure. Thus the frequencies of blocks picked out by those places where $z_n = 1$ are computed using product measure and this means that one gets the correct frequencies for the selected sequence along a subsequence (m_j) of the initial sequence (m'_j). Since this initial sequence was arbitrary, it follows that we get actual convergence to the correct frequencies and thus z defines an admissible selection rule.

The other direction of the theorem is more difficult and was first established by T. Kamae. One can combine these two results into a single one where one speaks of a finite automaton that has available a fixed, input-independent oracle. There are, of course, uncountably many distinct, completely deterministic sequences, e.g. for $\alpha > 1$, the sequences $\{m_j = [j\alpha]\}$. This means that a straightforward application of the usual measure-theoretic techniques cannot establish an almost-everywhere version of this last theorem.

References

1. R. Adler and B. Weiss, *Similarity of automorphisms of the torus*, Memoirs AMS **98**(1970).
2. R. Bowen, *Equilibrium states and the ergodic theory of Anosov diffeomorphism* SLN-470, 1975.
3. E. Glasner and B. Weiss, *Quasi-factors of zero entropy systems*, J. of AMS **8**(1995), reprint.
4. T. Kamae, *Subsequences of normal sequences*, Israel J. of Math. **16**(1973), 121–149.
5. T. Kamae and B. Weiss, *Normal numbers and selection rules*, Israel J. of Math. **21**(1975), 101–110.
6. M. van Lambalgen, *Random Sequences*, Thesis, Univ. of Amsterdam, 1987.
7. D. Ornstein, *Bernoulli shifts with the same entropy are isomorphic*, Adv. in Math **4**(1970), 337–352.
8. W. Parry, *Intrinsic Markov chains*, Trans. Amer. Math. Soc. **112**(1964), 55–66.
9. N. Sauer, *On the density of families of sets*, J. Combin. Theory Ser. A **13**(1972), 145–147.
10. S. Shelah, *A combinatorial problem; Stability and order for models and theories in infinitary languages*, Pacific J. Math. **41**(1972), 247–261.
11. Y. Sinai, *A weak isomorphism of transformations with an invariant measure*, Soviet Math. Doklady **3**)(1962), 1725–1729.

CHAPTER 9

Recurrence Rates and Entropy

Throughout this chapter $\{x_n\}$ will denote a finite, A-valued stationary stochastic process, $A = \{1, 2, \ldots a\}$. We have already mentioned the Shannon-McMillan theorem which gives a more combinatorial meaning to the entropy of the process. There is a "single orbit" refinement of this theorem that goes as follows:

THEOREM 9.1 (Shannon-McMillan-Breiman). *For an ergodic process $\{x_n\}$ and almost every realization $\{\xi_n\}$ we have that*

$$\lim_{n \to \infty} \frac{-1}{n} \log \mu([\xi_1 \ldots \xi_n]) = \text{entropy}(\{x_n\}).$$

Here $[\xi_1 \ldots \xi_n]$ denotes the cylinder set that consist of all the realizations of the process that agree with $\xi_1, \ldots \xi_n$ in the first n places. In one sense this is a single-orbit theorem, however, it asserts something concerning quantities which can only be defined if the entire statistics of the process are known. Our first topic will concern another meaning of the entropy – one which can be given an intrinsically single-orbit interpretation.

Let us begin with a description of the well-known Lempel-Ziv universal data compression scheme. Their goal was to give an algorithm for optimally compressing data that did not require one to know a priori what the statistical behavior of the data was. Shannon's fundamental coding theorems say that one can compress down to the entropy of the data and that this is the best that one can do. They gave an algorithm that achieved this optimal rate without any advance knowledge of

95

the statistics of the source. It is based on a **parsing** algorithm that divides the incoming data $\xi_1, \xi_2 \ldots \xi_n \ldots$ into blocks:

$$(\xi_1 \ldots \xi_{b(1)}), \ (\xi_{b(1)+1}, \ldots \xi_{b(2)}), \ldots$$

Each block is then encoded by a substitute which in general requires fewer bits of information to store. For simplicity we assume that the ξ_i's take only two values, $\{0, \ 1\}$. The definition of the blocks is inductive. The $k + 1$st block $(\xi_{b(k)+1}, \ldots \xi_{b(k+1)})$ is the shortest block beginning with $\xi_{b(k)+1}$ that does not appear earlier as one of the first k-blocks in our parsing. Thus for example the sequence

$$0\ 0\ 1\ 0\ 0\ 1\ 0\ 1\ 1\ 0\ 1\ 1\ 1\ 0\ 1\ 1\ 1\ 0\ 1\ 0$$

will be parsed into

$$(0)\ (0\ 1)\ (0\ 0)\ (1)\ (0\ 1\ 1)\ (0\ 1\ 1\ 1)\ (0\ 1\ 1\ 1\ 0)\ (1\ 0).$$

Notice that all but the last bit of the $k+1$-block already appear as one of the earlier blocks. The substitute for the $k + 1$-block in the compression scheme is the **index** of the earlier block which equals $(\xi_{b(k)+1} \ldots \xi_{b(k+1)-1})$ and the final bit $\xi_{b(k+1)}$.

They went on to show that for almost every realization of stationary ergodic process asymptotically the number of bits needed to compress $\xi_1 \ldots \xi_n$ is given by the entropy – i.e. $h(x) \cdot n$.

There are many variants of this basic parsing algorithm. It is in some ways more natural to look for the shortest block that begins with $\xi_{b(k)+1}$ that does not appear earlier in (ξ_i), not necessarily as one of the earlier blocks in the parsing. Now a simple index is not enough and one has to record the starting and ending places of the earlier location. In the analysis of all algorithms of this type a fundamental role is played by the **recurrence time**. This is defined as follows:

$$R_n((\xi_i)_1^\infty) = \inf\{r > n : \ \xi_{r+i} = \xi_i \text{ for all } 1 \leq i \leq n\}.$$

In words, R_n is the moment of the first reappearance, after time n, of the initial n-block of the observations $(\xi_i)_1^\infty$. A. Wyner and Y. Ziv [WZ-1989] drew attention to the importance of this recurrence time and proved half of the following theorem which was completed in [OW-1993].

THEOREM 9.2. *For a stationary ergodic process $\{x_n\}_1^\infty$ and a.e. realization $(\xi_i)_1^\infty$ one has that*

(1) $$\lim_{n \to \infty} \frac{1}{n} \log R_n((\xi_i)_1^\infty) = H(\{x_j\}_1^\infty).$$

As we have already remarked, such a result gives automatically a convergence result for all stationary processes – ergodic or not. In the non-ergodic case one cannot give easily the right-hand side in (1), one must simply say "the entropy of the ergodic component to which $(\xi_i)_1^\infty$ belongs. The formula (1) gives a meaning to the entropy which makes sense to an observer watching a simple output of a process about which he has **no** prior information. In this way it is much more of a single-orbit theorem than the usual SMB theorem. Before explaining the proof of this and other results related to the parsing algorithms of the type described above, I would like to give a heuristic explanation for it.

Let us recall a basic formula due to M. Kac. For an ergodic system $(X, \mathcal{B}, \mu, T)$ and a positive set $B \in \mathcal{B}$ let us define the first return time to B as:

$$r_B(x) = \inf\{n > 0: T^n x \in B\}.$$

FORMULA 9.3. *(M. Kac) For ergodic systems*

$$\int_B r_B (x) \, d\mu(x) = 1.$$

For fixed n, if B denotes the cylinder set defined by $\xi_1 \xi_2 \ldots \xi_n$ then Kac's formula says that, on average, $R_n((\xi)_1^\infty)$ behaves like $1/\mu(B)$ since we are now conditioning on the event that $x \in B$, and we should therefore be dividing by $\mu(B)$ in the formula above. On the other hand, the SMB theorem says that we expect $\mu(B)$ to equal, for large n, $\exp(-h\,n)$. Taken together these two facts suggest that theorem 9.2 should be true. A measure version of the theorem 9.2, which is related to it as is the weak law of large numbers to the strong law of large numbers, can be proved fairly easily along these lines, but the pointwise, single-orbit version requires more work.

Before continuing the main line I would like to pause and give a proof of Kac's formula which differs from the usual one. Not that this proof is any shorter, or easier, but I believe that it explains the result. Consider a typical point $x_0 \in X$, and the 1_B-name of its orbit, that is the sequence $\{1_B\,(T^i x_0)\}_{i=0}^{\infty}$. In this sequence

of zeros and ones, the ones represent the successive visits to B, and therefore if the

kth 1 is at place V_k it is clear that

$$r_B(T^{V_k} x_0) = V_{k+1} - V_k.$$

We can think of r_B as vanishing off of B and then if x_0 is generic for r_B, we see

immediately that the sum of r_B along the x_0 up to time n gives us exactly V_k

where k is such that $V_k \leq n < V_{k+1}$ and thus dividing by n we get in the limit 1 as

the formula demands. This way of picturing Kac's formula suggests analogues in

higher dimensions, for $\mathbb{Z}^d$-actions, where the notion of a first return is not so easy

to visualize.

Before explaining the proof of theorem 9.2, I would like to describe another

single-orbit version of the S-M theorem.

THEOREM 9.4. *Let $\{x_n\}_1^{\infty}$ be a stationary ergodic process with entropy equal
to h. For a.e. realization $(\xi_i)_1^{\infty}$ we have the following:*

For all $\epsilon > 0$ there is some N_ϵ, and for all $N \geq N_\epsilon$ if $M > \exp(hN)$ then

(i) *there is a collection of N-blocks $\mathcal{C}_N$ with $|\mathcal{C}_N| \leq \exp(h + \epsilon)N$ that covers a*
   $(1 - \epsilon)$-fraction of $\xi_1\,\xi_2 \ldots \xi_M$;

(ii) *no collection of N-blocks with fewer than $\exp(h - \epsilon)N$ elements can cover*
   even an ϵ-fraction of $\xi_1\,\xi_2 \ldots \xi_M$.

The discrete space $\xi_1 \ldots \xi_M$ is thus seen to have the property that the whole

probability space has in the usual S-M theorem. Notice too that the only restriction

placed on M is that it have a chance to see the typical blocks. If M is only on

the order of $\exp hN$ and no larger, then in (i) we can, of course, take the collection

of all N-blocks that appear. However, (ii) has content even in that case. To take

some of the mystery out of this result let me quickly say how to define the $\mathcal{C}_N$.

One chooses a large k so that $\frac{1}{k}\,H(x_1, \ldots x_k)$ is already very close to the

entropy of the process h. Then one takes an L to be large enough so that most

of the L-blocks have a k-block empirical distribution which is very close to the

true distribution. Finally, the N_ϵ should be taken large enough so that one sees

in $\xi_1\,\xi_2 \ldots \xi_M$, for $M > \exp hN_\epsilon$, the correct distribution of L-blocks. Now the collection $\mathcal{C}_N$ is taken to be those N-blocks that are $(1 - \frac{1}{10}\epsilon)$-covered by these good L-blocks. Matters have been arranged so that the blocks in $\mathcal{C}_N$ will cover $(1 - \epsilon)$ of $\xi_1 \ldots \xi_M$.

The estimate on the size of $\mathcal{C}_N$ comes about because it has a k-block empirical distribution which is close enough to the true distribution to ensure that its entropy, normalized by k, is only slightly larger than h. The final count is made much as we did in Chapter 7. Notice that the L is playing an auxiliary role – the $\mathcal{C}_N$ could be defined at once in terms of the k as being all those N-blocks whose empirical k-block distribution does not have a normalized entropy that exceeds h by very much.

The L-blocks are used to show that as soon as M is sufficiently large, these N-blocks will cover a $(1 - \epsilon)$-fraction of $\xi_1 \; \xi_2 \ldots \xi_M$.

The lower bound, (ii), is proved via a coding argument that I will explain first in connection with theorem 9.2 to which we now return.

PROOF OF THEOREM 9.2 – UPPER BOUND.

Let us define the upper limit of the recurrence time:

$$u(x) = \lim_{n \to \infty} \sup \; \frac{1}{n} \; \log R_n(x)$$

where $x = (\xi_i)_{-\infty}^{\infty}$, is now thought of as a point in the probability space on which the stochastic process is defined. We wish to establish the upper estimate that $u(x)$ is bounded from above by the entropy h.

Fix some $\delta > 0$, and for each n let $G_n \subset A^n$ be the set of n-blocks w of our process (whose state space we are denoting by A) such that

$$\mu([w]) \geq \exp(-n(h + \delta)).$$

According to the SMB theorem, for a.e. x we have that for n sufficiently large $\xi_1 \ldots \xi_n \in G_n$.

For any set B we have that the set

$$\{x : \; r_B(x) \geq L\}$$

has measure at most $1/L$, by Kac's formula. So if we take $u_0 > h + \delta$ then for any particular n for which $R_n(x) \geq u_0$ we know that the measure of that part of

$[\xi_1 \ldots \xi_n]$ satisfying this condition is at most

$$\exp(-n\, u_0).$$

Clearly the number of n-blocks in G_r is at most $\exp(n(h+\delta))$ so the set of x whose initial n-symbols belong to G_n and for which $R_n(x) > u_0$ is at most

$$\exp(-n\, u_0) \cdot \exp n(h+\delta) = \exp(-n(u_0 - h - \delta)).$$

These expressions for all n form a convergent series and so the classical Borel-Cantelli lemma says that $a.e.$ x can belong to only finitely many such sets. Since we know that $a.e.$ point is eventually in a G_n we conclude that $u(x) \leq u_0$. Since u_0 was an arbitrary number greater than h we are done with the proof of the upper bound. □

PROOF OF THEOREM 9.2 – LOWER BOUND

As before let us define

$$\ell(x) = \lim_{n\to\infty} \inf \frac{1}{n} R_n(x).$$

It is more convenient to work with constants rather than functions, so we remark that since

$$R_{n-1}(x) \leq R_n(T)$$

it follows from the ergodicity that $\ell(x)$ is equal to a constant, ℓ, $a.e.$ We will give a proof by contradiction. Assuming that $\ell < h$ we will show how to find a "small" collection of n-blocks that will cover most of the space – contradicting the ordinary S-M theorem. Choose some $\delta > 0$ so that $\ell + \delta < h$. For a large N_0 (whose size will be specified later) find an N_1 so that the set

$$D = \{x : \ R_n(x) \leq \exp(n(\ell+\delta)) \text{ for some } N_0 \leq n \leq N_1\}$$

has measure greater than $1 - \epsilon$ (for an $\epsilon > 0$ whose size will be specified below). These choices are possible by the definition of $\ell(x)$.

The set D being a set with measure at least $1 - \epsilon$, the ergodic theorem says that for m large enough, most points $x \in X$ will satisfy

$$\frac{1}{m} \sum_{D}^{m-1} 1_D(T^i x) \geq 1 - 2\epsilon.$$

Choose m much larger than N_1 and define disjoint blocks C_j as follows:

Let s_1 be the first integer such that $T^{s_1}x \in D$. For any point y in D we have associated an $n(y) \in [N_0, N_1]$ so that $R_{n(y)}(y) \leq \exp(n(y)(\ell + \delta))$. Define

$$C_1 = \xi_{s_1} \, \xi_{s_1+1} \cdots \xi_{s_1+n(T^{s_1}x)-1}.$$

Next s_2 is taken to be the least integer greater than $s_1 + n(T^{s_1}x)$ so that $T^{s_2}x \in D$, and then C_2 is defined as before, it covers the indices in $[s_2, \, s_2 + n(T^{s_2}x))$. We continue in this fashion defining blocks C_j, until we get to $m - \exp N_1(\ell + \delta)$. The key property that each of these blocks C_j has is that they repeat to the right of their actual occurrence in a place which is not more than $\exp n_j(\ell + \delta)$ away when n_j is the length of the block c_j. Furthermore, the fact that for most i, $T^i x \in D$ means that these blocks cover at least $(1 - 3\epsilon)$ of $(\xi_i)_1^m$.

Now we encode or enumerate these blocks in the following way:

(i) Fix the pattern of where these blocks occur – and their size. This requires marking the initial and final indices in each block and since each block is at least N_0 we have fewer than m/N_0 initial and final markers. This will contribute a small exponential number if N_0 is large enough.

(ii) For each block C_j we record only the distance to its first recurrence to the right.

(iii) Any place not covered by a block we record the actual symbol ξ_i that appears there.

It is now easy to see that with this data the $(\xi_i)_1^m$ can be uniquely recovered by filling in the C_j's from right to left – i.e. beginning with the last C_j that was defined. It is also straightforward to calculate the total number of different encoded words that can be obtained in this way. The decisive contribution comes from (ii) and because $\ell + \delta < h$ we get a definite exponential drop in the total number of different words that have this type of repetition.

It is now clear that we have contradicted the S-M theorem and the proof is complete.

We can come back now to the single-orbit version of the S-M theorem, theorem 9.4, and explain how to prove the second part. The assertion is that for the typical $(\xi_i)_1^\infty$, one cannot cover $(\xi_i)_1^M$ by a small collection of N-blocks. To prove this we

assume the contrary, namely that for a set of positive measure, say $\alpha > 0$, of x's there are infinitely large N's for which such an efficient covering is possible. One uses the same kind of argument that we just used to cover say $\alpha/2$ of most names by these blocks that are covered by small collections of N-blocks.

There are two new points in the proof here. The first is the fact that there are many different small collections. However, when we count how many, it turns out that this is a small exponential in M. Indeed the total number of subcollections of A^N of cardinality at most $\exp N(h - \epsilon)$ is certainly no greater than

$$|A^N|^{\exp N(h-\epsilon)}$$

which is dominated by $\exp([(\log A) \cdot N \exp(-N\epsilon)] \cdot M)$ since $\exp Nh < M$. For fixed ϵ the coefficient of M clearly tends to zero as N tends to ∞ and so this factor is negligible.

The second new point is that only a small (but fixed) fraction of the M-name is covered by this small collection of names. However, using the first part of the theorem that we have already explained, one can cover most of the rest of the M-block by blocks from a collection numbering no more than $\exp N(H+\epsilon)$ elements so that the drop on the fixed $\alpha/2$ fraction will bring the total count down to below the entropy h and thus we can complete the proof as before. The count of patterns and how one fills in what is in the small ϵM fraction that are in none of the N-blocks we just discussed, is done as before.

Let us go back now to the Lempel-Ziv algorithm and see what its relationship to these results is. With arguments very similar to what I have already explained one can prove, for example, the following result which gives a fairly precise description of the size of the blocks in the Lempel-Ziv parsing.

THEOREM 9.5. *For a finite-valued ergodic stationary stochastic process* $\{x_n\}$ *and for a.e. realization* $(\xi_i)_1^\infty$ *we have that for all* $\epsilon > 0$, *if n is sufficiently large then a* $(1 - \epsilon)$-*fraction of* $(\xi_i)_1^n$ *is covered by blocks in the Lempel-Ziv parsing with size in the interval*

$$\left[\frac{1-\epsilon}{h} \log n, \ \frac{1+\epsilon}{h} \log n\right].$$

This sharp result gives a sharp estimate on the number of blocks possible in the Lempel-Ziv parsing and gives immediately the result that the compression rate of this algorithm is asymptotically optimal for any ergodic process. We can recapitulate the theme of this chapter in the following way. The usual definition of the entropy of a stochastic process requires one to know the process fully, all of its finite distributions, in order to compute it. In principle, of course, observing a single typical orbit of the process out to infinity will enable us to compute all of these finite distributions and then the entropy. One can ask, however, for an "on line" scheme which will calculate better and better estimates for the entropy as more and more observations ξ_i are made. The SMB theorem, in spite of its being a single-orbit theorem, does not do this. In contrast we have seen here three different ways of doing so since each of the theorems 9.2, 9.4 and 9.5 can be used as "on line" estimates for the entropy. Of the three the simplest one, conceptually, is perhaps 9.2. The one closest to the original entropy definition is 9.4 while 9.5 probably leads to the most efficient way of calculating the entropy. In the next chapter we shall study some more problems of this type, that is how to answer various questions concerning a process by observing more and more outcomes of a single stream of data.

References

1. A. Lempel and J. Ziv, *Compression of individual sequences via variable rate coding*, IEEE Trans. Inform. Theory, **24**(1978), pp. 530–536.
2. A. Lempel and J. Ziv, *A universal algorithm for sequential data compression*, IEEE Trans. Inform. Theory, **23**(1977), pp. 337–343.
3. D. Ornstein and B. Weiss, *How sampling reveals a process*, Ann. Probab. Theory, **18**(199), pp. 905–930.
4. D. Ornstein and B. Weiss, *Entropy and data compression schemes*, IEEE Trans. on Info. Theory, **39**(1993), pp. 78–83.
5. A. Wyner and J. Ziv, *Some asymptotic properties of the entropy of a stationary ergodic data source with applications to data compression*, IEEE Trans. Inform. Theory, **35**(1989), pp. 1250–1258.
6. J. Ziv, *Coding theorems for individual sequences*, IEEE Trans. Inform. Theory, **24**(1978), pp. 405–412.

CHAPTER 10

Universal Schemes

The recurrence rate result of the previous chapter and the Lempel-Ziv data compression scheme were examples of universal schemes. That is to say they were methods for extracting some kind of information out of a simple sample sequence of a process without knowledge of what the process is. Not every question that one can formulate in this way has an affirmative answer. Let us contrast two superficially similar basic prediction questions.

QUESTION 10.1. *Suppose an observer is given the successive outputs of a stationary stochastic process $\{\xi_n\}_0^\infty$. After observing ξ_0^n the observer is required to give his estimate for the conditional distribution of ξ_{n+1} given ξ_0^n. Can one describe such a function $\mathcal{D}_n(\xi_0^n)$ such that for every ergodic process and almost every realization one has*

$$\lim_{n \to \infty} \|\mathcal{D}_n(\xi_0^n) - \text{Distribution } (\xi_{n+1}|\xi_0^n)\|_1 = 0?$$

The point is that the observer does not know what the statistics of the underlying process are, he merely observes ξ_0^n and is called upon to then estimate the conditional distribution of ξ_{n+1}.

QUESTION 10.2. *Suppose an observer is given more and more of the past outcomes of a process $\{\xi_n\}$, i.e. ξ_{-n}^{-1}, can he then give an estimate $\mathcal{E}_n$ for the conditional distribution of ξ_0 given the past ξ_{-n}^{-1} that will satisfy*

$$\lim_{n \to \infty} \|\mathcal{E}_n(\xi_{-n}^{-1}) - \text{Dist}(\xi_0|\xi_{-\infty}^{-1})\|_1 = 0$$

for almost every realization of every ergodic process?

For questions in the mean, it seems clear that there is no difference between these two questions because of the stationarity. However, for single-orbit questions there is a striking difference between the two. For the second question, D. Ornstein in [O-1978] gave an affirmative answer, whereas his student D. Bailey, in his thesis [B-1976] showed that the first question has a negative answer. In other words any scheme $\mathcal{D}_n$ can be fooled into giving the wrong answer infinitely often for a set of positive measure of some stationary ergodic process. A very elegant scheme for question 2, somewhat simpler than Ornstein's, has been given recently by G. Morvai, S. Yakowitz and L. Györfi and that will be our first objective in this chapter. We will then describe briefly a construction of a family of processes that shows why there is no universal scheme which will work for question 1.

But first we will turn to more general questions concerning the determination of the entire underlying process from the observation of a single output sequence. Here it will turn out that a rather complete answer can be given, not for all stationary processes, but for the chaotic ones. We will find here a good example of how randomness, in some paradoxical way, enables one to make some kind of definite statements about whole processes. Before formulating the question we need to extend the definition of the $\bar{d}$-distance that we used in Chapter 4 from sequences to processes. Given two processes $\{x_n\}$, $\{y_n\}$ with the same state space A by a **joining** we mean a shift-invariant measure on $(A \times A)^{\mathbb{Z}}$ whose two coordinate projections have the distribution of the $\{x_n\}$ and $\{y_n\}$ processes respectively. Thus for any joining we can make sense of events like $\{x_0 = y_0\}$. The $\bar{d}$-distance between processes is

$$\bar{d}(\{x_n\}, \{y_n\}) = \inf\{P(x_0 \neq y_0) : \text{ all stationary joinings of } \{x_n\}, \{y_n\}\}.$$

For ergodic processes it is not hard to see that this is the same as the infimum over all generic points $\{\xi_n\}$, $\{\eta_n\}$ for the two processes of $\bar{d}\,((\xi_n), (\eta_n))$, where this $\bar{d}$-distance is the metric on names given by:

$$\varlimsup_{n \to \infty} \frac{1}{n} |\{0 \leq j < n : \xi_j \neq \eta_j\}|.$$

QUESTION 10.3. *Can one find mappings $\mathcal{G}_n$ from observations ξ_0^n of the outputs of stationary processes to stationary processes themselves so that for every ergodic process $\{x_n\}$ and almost every realization $\{\xi_n\}$ we have*

$$\lim_{n \to \infty} \bar{d}\,(\mathcal{G}_n(\xi_0^n),\ \{x_k\}) = 0.$$

What we want is that after observing a finite number of outputs, one make some educated guess as to the nature of the process and we want that this guess be close in the $\bar{d}$-metric to the process. It is perhaps not so surprising that in general the question has a negative answer, one cannot find such a universal guessing scheme. However, if one is dealing with the chaotic systems that exhibit maximum randomness - by which I mean the independent processes and processes obtained from their factor mappings - then the answer is yes!

The nonexistence of **truly** universal schemes extends even to far simpler questions such as the discrimination between processes. Discrimination is the following question:

QUESTION 10.4. *Can one find functions $\mathcal{D}_n$ from pairs of observations (ξ_0^n, η_0^n) to the interval $[0,1]$ such that $\mathcal{D}_n(\xi_0^n, \eta_0^n)$ converges to zero if and only if the ξ_0^n, η_0^n are typical samples from the same ergodic process?*

For this question too the answer is — no. Namely there are no functions $\mathcal{D}_n$ which have the above property for all pairs of ergodic stationary processes.

Having formulated the questions we go back and explain the answers in more detail. We begin with question 10.2. We read more and more of the past of a sequence of observations ξ_j, $j < 0$ and define inductively words w_i and indices $j(i-1)$ as follows:

$$w_1 = \xi_{-1}, \ j(0) = 1$$

$$\vdots$$

$$w_{k+1} = \xi_{-j(k)}^{-1} \text{ where } j(k) \text{ is the first time that block } w_k \text{ recurs,}$$

$$\text{i.e. } \xi_{-j(k)}^{-j(k)+j(k-1)-1} = w_k \text{ with } j(k) > j(k-1)$$

$$\vdots$$

Since for a.e. realization this will happen for all k, we do not bother to say how to define w_{k+1} if w_k never recurs. As soon as we are able to determine $w_1, w_2, \ldots w_{k+1}$ our estimate for the conditional distribution of x_0 given the past $\xi_{-j(k-1)}^{-1}$ becomes the empirical distribution of the symbol following these first reappearances of w_i, $1 \leq i \leq k$. To be precise, for a particular symbol a, the estimate for

$P(x_0 = a| \xi_{-m}, \ m < 0)$ becomes

$$\frac{1}{k} \, | \, \{2 \leq \ell \leq k+1 : \ \xi_{-j(\ell-1)+j(\ell-2)} = a\}.$$

THEOREM 10.5. *The empirical distributions described above converges to the conditional distribution* $P(x_0|\xi_{-m}, \ m \geq 1)$ *for almost every realization of an ergodic finite valued stochastic process.*

The well-known martingale convergence theorem implies that for almost every realization, if we calculate the finite conditional distributions

$$P(x_0 = a|\xi_{-n}^{-1})$$

then they will converge to the true $P(x_0 = a|\xi_m, \ m < 0)$. This is clearly what we would like to do, however, we do not know the statistics of the process. Nevertheless, for the proof, we can imagine that they are known and then we can define random variables.

$$Z_k = 1_{\{\xi_{-j(k)+j(k-1)}=a\}} - P(x_0 = a|w_k).$$

The theorem is equivalent to the assertion that almost surely

$$\lim_{N\to\infty} \frac{1}{N} \sum_{1}^{N} Z_k = 0.$$

This is exactly what we get from the martingale convergence theorem, and we sketch now how to prove this.

For a fixed value of w_k one sees, by examining the successive occurrences of w_k, that the average of Z_k is zero when taken over a long portion of the orbit. Next we want to show that for $\ell > k$ the variables Z_k and Z_ℓ are uncorrelated, that is

$$\int Z_k \, Z_\ell \, dP = 0.$$

For this one conditions on the event that $\xi_{-j(k)+j(k-1)} = a$ and then observes that this simply determines a certain class of possible w_ℓ's. But for each value of w_k we see that

$$\int_{[w_k]} Z_\ell \, dP = 0$$

and so the Z_ℓ's are uncorrelated.

At this point we can apply a very general convergence theorem that applies to any orthogonal or uncorrelated sequence of random variables. This result, although quite old, is not as widely known as it should be, and therefore a proof is included for the case at hand - when the variables are bounded.

PROPOSITION 10.6. *If* $\{u_n\}_1^\infty$ *are uniformly bounded functions on a finite measure space* $(X,\ \mathcal{B},\ \mu)$, *with zero integral and satisfy*

$$\int_X u_n(x)\, u_m(x)\, d\mu(x) = 0 \qquad \text{all } n \neq m$$

then for $\mu - a.e.\ x$ *we have*

$$\lim_{n\to\infty} \frac{1}{n} \sum_1^n u_i(x) = 0.$$

PROOF. We apply the same device H. Weyl did in his famous paper on equidistribution mod 1 and consider the sums along the subsequence of squares:

$$s_{n^2} = \frac{1}{n^2} \sum_1^{n^2} u_i.$$

A trivial computation reveals that the sum

$$\sum_1^\infty \int_X |s_{n^2}(x)|^2\, d\mu(x) < +\infty$$

whence $s_{n^2}(x)$ converges to zero μ-almost everywhere. Finally the assumption that the u_i's are uniformly bounded gives a uniform estimate for $n^2 \leq k < (n+1)^2$ of

$$|s_{n^2}(x) - s_k(x)| \leq \left| \frac{1}{n^2} \sum_1^{n^2} u_i(x) - \frac{1}{k} \sum_1^{n^2} u_i(x) \right| + \frac{1}{k} \left| \sum_{n^2+1}^k u_i(x) \right|$$

$$\leq M \cdot \frac{(k - n^2) \cdot n^2}{k \cdot n^2} + \frac{k - n^2}{k} \cdot M \leq \frac{5M}{n}$$

where M is the uniform bound on the $u_i(x)$'s. Thus the convergence to zero along the subsequence implies the convergence along the entire sequence of averages. □

Putting all of the above together we have established theorem 10.5. The algorithm described above, while quite simple, is not very efficient in that most of the information in ξ_{-m}^{-1} that may bear on the relevant conditional probability that we are trying to estimate is going to waste. The more intuitive algorithm of fixing some length L and then examining **all** occurrences of ξ_{-L}^{-1} in ξ_{-m}^{-1} to determine an empirical distribution uses much more information but since the process is unknown we have no idea how to choose L as a function of m so as to get convergence for **all** processes.

We turn now to question 10.1. Here the fact that the block we are conditioning on is constantly changing means that we cannot simply use the preceding result. In fact it is not hard to see that any given estimation scheme can be foiled infinitely often by some process (with probability one). The idea is to construct a family of processes with varying distributions so that there are rare events, E_n, which a typical point will encounter infinitely often but these rare events could have different conditional probabilities associated with the next output depending upon which process of the family we were looking at. This was Bailey's method. Later on B. Ryabko described a simpler family and the following example is based on his. We suppose then, that $\mathcal{P}_n$ is a prediction scheme for stationary stochastic processes with three symbols $\{0,\ 1,\ 2\}$ that produces, for each input ξ_1^n, a distribution on x_{n+1} with the property that for any ergodic stationary process with probability one we have

$$\lim_{n\to\infty} \|\mathcal{P}_n(\xi_1^n) - \mathrm{Dist}(x_{n+1}|\xi_1^n)\|_{\ell^1} = 0.$$

We proceed to describe a family of stochastic processes indexed by sequences $\omega \in \{1,2\}^{\mathbb{N}}$. Begin with an independent process of fair coin tossing with outcomes $\{H,T\}$ equally likely. Now replace each occurrence of an H by a 0, and each maximal string of consecutive T's of length m by the sequence $\omega(1)\,\omega(2)\ldots\omega(m)$. Note that, abstractly, all of these processes are simply isomorphic to the original i.i.d. Now to construct the confounding process, begin by defining stopping times $\tau_1 < \tau_2 < \ldots$ on the i.i.d. process where τ_k is the first time that we have seen k-consecutive T's. Start with the process where $\omega(i) = 1$ for all i. There is some k, large enough, so that with probability .99, the scheme $\mathcal{P}$ will correctly predict the conditional probability of x_{τ_k+1} given the observations up to that point to within an accuracy of .99. We now switch and consider

$$\omega^{(2)}(i) = \begin{cases} 1 & , \quad i \le k \\ 2 & , \quad k < i. \end{cases}$$

Clearly for the $\omega^{(2)}$-process the prediction scheme will err on a set of probability at least .99 by a significant amount. Now, since the scheme is universal, it will eventually correctly predict the ω^2-conditional distribution on the whole space and thus for k_2 large enough it will predict $x_{\tau_{k_2}+1}$ very well. We then define

$$\omega^{(3)}(i) = \begin{cases} \omega^{(2)}(i) & i \leq k_2 \\ 1 & k_2 < i \end{cases}$$

The procedure continues ad infinitum and a limiting sequence is constructed which with probability one will infinitely often give an incorrect prediction.

The problem disappears if one is satisfied with an average result. In that case the usual ergodic theorem can turn a scheme which works well on the past in the mean - to an almost everywhere result - on average. The details are pretty straightforward and we shall not enter into them.

We turn now to question 10.3. The key to the positive answer for the class of chaotic processes lies in their characterization, given by D. Ornstein in his pioneering work on Bernoulli processes. They key definition is the following:

DEFINITION 10.7. A stationary process $\{x_n\}$ is said to be **finitely determined** if for any $\epsilon > 0$, there are $\delta > 0$ and m_0 such that any stationary process $\{y_n\}$ satisfying

(1) $$|\text{entropy}(x_n) - \text{entropy}(y_n)| < \delta$$

(2) $$\|\text{Dist}\{x_i\}_1^{m_0} - \text{Dist}\{y_i\}_1^{m_0}\|_{\ell_1} < \delta$$

also satisfies

(3) $$\bar{d}(\{x_n\}_1^\infty, \{y_n\}_1^\infty) < \epsilon.$$

It is clear that condition (1) is necessary, since finite distribution alone cannot guarantee long-range behavior. The point is that given (2) the entropy alone is enough to determine the long-range behavior of the process. It is not obvious that there are any processes at all that satisfy this definition. We cannot enter into the whole story here. Suffice it to say that one can show that i.i.d. processes satisfy this property, i.e. are finitely determined, as are any processes codeable from i.i.d. processes, and that this characterizes them. It is a further theorem of Ornstein that such processes are classified up to isomorphism by the entropy alone. For the details of all this, and much more, see for example [O-1974]. With this concept at our disposal it should become clear that to answer question 10.3 in the affirmative for the class of finitely determined processes what we need to do is estimate, with greater and greater accuracy, both the finite distributions and the entropy and then offer as our guess for that process that produced the sequence ξ_1^n any process with that entropy and finite distributions. We have already seen several schemes for estimating the entropy in Chapter 7 and Chapter 9 while the basic method for estimating finite distributions is simply to use the empirical distribution of small blocks in ξ_1^n. To be sure one still has to decide up to what size k to take these blocks and then one still has to choose a process with the correct entropy and with that finite distribution. It turns out that a rather simple and robust bootstrapping technique will work here.

Here is an informal description of how to construct a good guess for the process that generates the string ξ_1^n. Fix some sequence $L(k)$ that grows faster than any exponential, for example $L(k) = k^k$. For all n in the interval

$$L(k) \leq n < L(k+1)$$

do the following: Choose k-blocks in ξ_1^n uniformly over all positions $1, 2, \ldots n - k$ and independently and concatenate them. That is to say that our process consists of independently chosen k-blocks from the observed string ξ_1^n. It is clear that the resulting processes converge in finite distribution to the ergodic process that generated ξ_1^∞ for almost every realization ξ_1^∞. The fact that the entropy converges to the entropy of the independently concatenated blocks converges to the true entropy can be proved based on the results of Chapter 7. For details see [OW-1990]. The sequence $L(k)$ was chosen to be a superexponential since we wanted to give a scheme which will work for all finite valued processes. If one knows an a priori upper bound for the entropy of the process, for example if one knows that the alphabet of the process has no more than $L - 1$ elements, then one can take for $L(k)$ simply L^k.

The scheme is robust in the sense that if it is applied to a sample sequence $\widehat{\xi}_1^\infty$ that is within ϵ in $\bar{d}$ of a typical sequence ξ_1^∞ that was produced by a process $\{x_n\}$ then eventually the process that the scheme produces will be within ϵ, in $\bar{d}$, of the true process. Thus even if we receive the observations degraded by some errors, we can still do as well as expected - namely get a process within that degree of error to the correct process. With this positive answer to question 10.3 for the class of finitely determined processes we can also give a positive answer to question 10.4 in the class of chaotic processes. This was the question of deciding whether two sequences ξ_1^∞, η_1^∞ come from the same or from different processes. All that needs to be done is to construct one of our estimates for each of ξ_1^n and η_1^n and then compute the $\bar{d}$-distance between them and that distance will converge to the true $\bar{d}$-distance between the processes that generated ξ_i and η_i respectively. In particular it will be zero if and only if the sample sequences were drawn from the same process.

The negative answer to the same question for the entire class of processes requires a construction of a type that we have not discussed in these notes, even though it is quite standard. Once again the interested reader is referred to the paper [OW-1990].

I would like to conclude by pointing out that our discussions throughout this chapter rely in an essential way on the stationarity of the process. For nonstationary processes one does not have the notion of generic points and the supports of distinct processes can be intertwined in a very complex way. Consider, for example, two classes of processes on $\{0,1\}^{\mathbb{N}}$:

$$\mathcal{P}_0 = \{\text{processes } \mu \text{ for which } \lim_{n \to \infty} \mu(x_n = 0) = 1\}$$
$$\mathcal{P}_1 = \{\text{processes } \mu \text{ for which } \lim_{n \to \infty} \mu(x_n = 1) = 1\}.$$

For any two fixed measures $\mu_0 \in \mathcal{P}_0$, $\mu_1 \in \mathcal{P}$ it is easy to see that there are disjoint Borel sets A_0, A_1 that support μ_0, μ_1 respectively. Thus there is a nice function of the entire sequence of observation $\{\xi_i\}$ that distinguishes between μ_0 and μ_1. However, D. Blackwell has shown that there is no Borel function that will distinguish between the classes $\mathcal{P}_0$ and $\mathcal{P}_1$. For stationary processes all such questions where one is allowed to look at the entire sequence $\{\xi_i : 1 \leq i < \infty\}$ can be easily answered since distinct ergodic measures have distinct generic points and thus disjoint supports.

References

1. D. Bailey, *Sequential schemes for classifying and predicting ergodic processes*, Ph.D. dissertation, Stanford Univ.
2. D. Blackwell, *There are no Borel SPLITs*, Ann. Probab. **8**(1980), 1189–1190.
3. G. Morvai, S. Yakowitz, and G. László, *Nonparametric inference for ergodic, stationary time series*, Ann. Statist. **24**(1996), no. 1, 370–379.
4. D. Ornstein, *Ergodic Theory, Randomness and Dynamical Systems*, Yale Univ. Press, New Haven, Conn, 1974.
5. D. Ornstein *Guessing the next output of a stationary process*, Israel J. Math. **30**(1978), 292–296.
6. D. Ornstein and B. Weiss, *How sampling reveals a process*, Ann. Probab. **18**(1990), 905–930.
7. B. Ya. Ryabko, *Prediction of random sequences and universal coding*, Problems Inform. Transmission **24**(1988), no. 2, 87–96.

Selected Titles in This Series

(*Continued from the front of this publication*)

For a complete list of titles in this series, visit the
AMS Bookstore at **www.ams.org/bookstore/**.

DATE DUE